© Laurie Lane Studios

DOUGLAS J. EMLEN is a professor of biology at the University of Montana. He is the recipient of the Presidential Early Career Award for Scientists and Engineers from the Office of Science and Technology Policy at the White House; multiple research awards from the National Science Foundation, including its five-year CAREER award; a Young Investigator Award; and the American Society of Naturalists' E. O. Wilson Naturalist Award. His research has been featured in outlets including *The New York Times* and NPR's *Fresh Air*.

DAVID J. TUSS is a graphic artist who specializes in blending technical accuracy with vivid, lifelike compositions. His work is featured in textbooks, scientific articles, and technical biology papers. He lives in Helena, Montana, where he works as a wilderness ranger, natural science illustrator, and public school science and art teacher.

Also by Douglas J. Emlen

Evolution: Making Sense of Life
(with Carl Zimmer)

A Handbook of Biological Investigation
(with Harrison W. Ambrose iii, Katharine Peckham Ambrose, and Kerry L. Bright)

Praise for

Animal Weapons

"One of our leading evolutionary biologists, Doug Emlen, delves into the deep meaning of entities . . . to take the reader on a joyous ride of discovery about nature and the human experience. *Animal Weapons* is an authoritative, knowledgeable, and epic narrative of one of the dominant themes of life on earth, including our own. Emlen's curiosity, passion, and storytelling prowess make this magisterial little volume leap from the page." —Neil Shubin, author of *Your Inner Fish*

"Emlen's excellent writing will draw in readers intrigued by astonishingly powerful weapons, both in the wild and in the military, and how they have evolved owing to selective pressures."

—*Library Journal* (starred review)

"A skilled storyteller, Emlen takes the reader to Panama to observe dueling Harlequin beetles; to Tanzania to collect elephant excrement—and other dung beetles that live on it—and to tropical forests, where stalk-eyed flies defend their harems of females. The weapons discussed in these stories are illustrated in stunning color plates." —*Science*

"In this original study, Emlen tours offensive and defensive anatomy and behaviors across aeons and taxa, from Tyrannosaurus rex's fearsome teeth to ibex horns and amphibian poisons. He sharpens the discussion by interweaving parallels with humanity's own evolving arsenal."

—*Nature*

"Absorbing . . . Throughout the book, Emlen's demonstrations of the many parallels between human and animal weapons are fascinating, even when the possibilities are frightening. . . . Emlen is not a hurried or simplistic storyteller. He is a writer of nuance, and he traveled to many different environments to get the story." —*Kirkus Reviews*

"Emlen infuses scientific explanations with entertaining anecdotes from his field research at the University of Montana. Each step of the way, he provides parallels with human weapon development and design—from ancient civilizations to weapons of mass destruction—and the evolutionary process of animals. While his conclusions about the human arms race are dire, it is his description of animal weaponry in action and in evolution that will captivate." —*Publishers Weekly*

"What's so appealing about this book is the way in which Dr. Emlen seamlessly weaves in his personal and extensive field experiences studying among other animals, dung beetles, who also display formidable weapons, and solid scientific research. Readers don't have to know much about animal behavior or evolutionary theory to come away with an understanding of why many different animals have invested a good deal of energy in evolving extreme weapons in their own arms race that might also compromise them in contexts other than battle."

—*Psychology Today*

"Dramatically illustrated, *Animal Weapons* is a fascinating look at the extremes of animal appendages and behaviors. Whether focused on the fiddler crab's enormous claw or the rhinoceros beetle's lethal horns, Emlen tells stories of natural selection that rival *Animal Planet*'s finest." —*Shelf Awareness*

"This is a great read not only for the stories of conflicts and weaponry in a diversity of animals, but also for the history of human weaponry, and the highly relevant message about arms races the author reads from both." —Bernd Heinrich, author of *Winter World* and *Why We Run*

"*Animal Weapons* is a must-read, especially those of us who are interested and concerned about human weapons development. As Douglas Emlen shows definitively, arms races are not something we as a species invented, but instead the most natural thing in the world."

—Robert L. O'Connell, author of *The Ghosts of Cannae* and *Fierce Patriot*

"Doug Emlen has done a superb job of bringing together the stories of animal and human weapons. He makes the biology behind the evolution of weapons understandable for this soldier and engineer and convincingly illustrates the human animal's problems in controlling or avoiding catastrophe in the age of weapons of mass destruction."

—Lieutenant General John Myers

"From antlers and tusks to claws and horns, University of Montana biologist Douglas Emlen investigates how animals have grown in different ways to defend themselves. In addition to outlining the development and stories about animal weapons, Emlen compares this to the development of human weapons. Just as the claws or antlers of animals may grow to proportions too large to be supported, humans are letting their own weapons get wildly out of control. A fascinating read."

—*Missoulian*

"While Emlen spins a dramatic tale with periodic firsthand accounts of some of his own scientific adventures and discoveries, Tuss grabs the reader's eye with his powerful art. . . . All of the dramatic drawings are based on the most accurate scientific information Tuss could gather, right down to holding the skull of a saber-toothed cat in his hands to study the position of the eyes." —*Independent Record* (Helena)

"A tour de force that parallels wildlife evolution of offensive and defensive weapons with our own social development of armed combat . . . The parallels are amazing. . . . Read this book and then pass your copy on to a politician who might make a difference."

—*The Buffalo News*

"Well worth reading. It is an absorbing look at the development of physical characteristics in nature that serve to drive forward the strongest of the species, those equipped with the resources to be able to develop and maintain features that come with a cost to their wearer. . . . Throughout the book are beautiful illustrations by artist David Tuss that truly bring the text alive." —*Staten Island Live*

"*Animal Weapons* offers dozens of small lessons on the intricacies of life on earth and a bigger one about how we forget that humans are just animals, too. It's a pleasure to read and packed with interesting tidbits, whether your interests swing toward military history or naturalist trivia."

—*Missoula Independent*

"Long story short, [Emlen] may have stumbled onto a new way to look at evolution after poking through dried poop in East Africa."

—*Vice*

ANIMAL WEAPONS

THE EVOLUTION OF BATTLE

Douglas J. Emlen

Illustrated by David J. Tuss

Picador Henry Holt and Company New York

Printed in the United States of America. For information, address
Picador, 175 Fifth Avenue, New York, N.Y. 10010.

picadorusa.com
twitter.com/picadorusa • facebook.com/picadorusa
picadorbookroom.tumblr.com

Picador® is a U.S. registered trademark and is used by
Henry Holt and Company under license from Pan Books Limited.

For book club information, please visit facebook.com/picadorbookclub
or e-mail marketing@picadorusa.com.

Designed by Meryl Sussman Levavi

The Library of Congress has cataloged the Henry Holt edition as follows:

Emlen, Douglas John, 1967–
Animal weapons : the evolution of battle / by Douglas J. Emlen; illustrated by David J. Tuss.—First edition.
p. cm.
Includes bibliographical references and index.
ISBN 978-0-8050-9450-3 (hardcover)
ISBN 978-1-4299-4739-8 (e-book)
1. Animal weapons. 2. Animal defenses. 3. Defensive (Military science). I. Title.
QL940.E45 2014
591.47—dc23 2014004772

Picador Paperback ISBN 978-1-250-07531-4

First published by Henry Holt and Company, LLC

First Picador Edition: December 2015

10 9 8 7 6 5 4 3 2 1

I know not with what weapons World War III will be fought, but World War IV will be fought with sticks and stones.

—Albert Einstein

Contents

Preface

For as long as I can remember I've been obsessed with big weapons, which is rather surprising given that I descend from a long line of Quakers. On field trips to the natural history museum it wasn't birds or zebras that caught my eye; it was mastodons with curling tusks, or triceratops with five-foot-long horns. In every room, it seemed, loomed another species with a crazy protrusion jutting from its head, or from between the shoulder blades, or from the end of its tail. Gallic moose wielded twelve-foot-wide antlers, and arsinotheres had horns six feet long and a foot wide at the base. I couldn't peel my eyes from these creatures. Why were their weapons so big?

As I grew and, in particular, as I learned more about biology, I realized that "big" had little to do with absolute size. Extreme weapons were all about proportion. Some of the most magnificent structures are borne by tiny creatures. Hiding in drawer after drawer of dried, pinned specimens in museum archives, for example, are uncountable numbers of oddball species: beetles with front legs so long they have to be folded awkwardly around the animal in order to shut the lid on the case, or horns so big the animals have to be mounted in the drawer sideways. Many species are so small that their weapons become apparent only with a microscope: twisted tusks protruding from the faces of West African wasps, for example, or broad, branched antlers adorning the faces of flies.

I began my career determined to study extreme weapons, so I set out to find the craziest, most bizarre animals that I could. I also wanted my research to take me someplace exotic. In my case, this meant the tropics, so I narrowed my search. My study animals needed to be easy to find in large numbers, to observe in the wild, and to rear in captivity. As fate would have it, the animals that best fit this bill were dung beetles. I resisted at first. After all, dung beetles lack the panache of elk or moose and, well, they eat dung. Dung beetles were also a tough sell whenever I tried to explain what I did to anyone outside of biology. My father-in-law springs to mind—he's a retired U.S. Air Force colonel—and I'll never forget breaking it to him that I wanted to take his daughter with me to a remote field station in the thick of a tropical rainforest so I could watch dung beetles.

But dung beetles really were the best animals for testing the ideas I wanted to test, and there were lots of them in the tropics. Squat like little tortoises, these beetles were armed to the teeth with spectacular horns. Better yet, almost nothing was known about how these weapons were used, why they were so big, or why species differed so incredibly in the numbers and shapes of horns that they produced. To a biologist, that kind of unknown is intoxicating. Like exploring the depths of the ocean or outer space, I was going to plunge into the abyss of the biological void, and I was going to learn about extreme weapons in the process.

Two decades later, I remain just as awestruck by beetle weapons as I was that first year in the tropics. I've followed their stories to Africa, Australia, and throughout Central and South America. I've also had a chance now to step back from beetles, to consolidate the lessons learned by biologists studying a plethora of extreme weapons in animals ranging from moose flies and fiddler crabs to elephants and elk. These are the stories I set out to tell in these pages, weaving together for the first time the narratives of nature's most extravagant creatures.

In the process of weaving their histories together, it became clear that there was another species that belonged in the mix: humans. The more I sought common threads—themes uniting the stories of diverse animal species—the more apparent it became that these threads applied to our own weapons, too. In the end, my book about animal weapons evolved into a book about extreme weapons everywhere. I pored deeper into the literature surrounding our past, searching for the environments and circumstances in which our most elaborate weapons evolved. To my amazement, these circumstances truly were the same, and I realized I couldn't tell one story without telling the other. Back and forth I went, as the biology of animal weapons and our weapons fused, inextricably woven into a single tale. This is a book about extreme weapons. Let's just leave it at that.

ANIMAL WEAPONS

Extremes

It was a cold, clear mountain night. The Milky Way streaked across the sky; jagged peaks loomed black against the starlight. A college buddy of mine and I were camping in Rocky Mountain National Park. It was early fall—peak of the rut for elk. I'd insisted we take the most remote campsite possible, and we'd pitched our tent as far away from the rest as we thought we could get away with. I wanted to be surrounded by aspen and cottonwood, not other people.

Somewhere around two a.m. I jolted awake, sleep still spinning in my head. A gunshot? I sat there in the silence listening. It came again—*crack*! In an instant I knew what was happening, and it was no gunshot. I shook Scott awake and we bolted from the tent. Almost a ton of testosterone-driven rage exploded beside us in the blackness not twenty feet away. Mature bull elk easily weigh in at eight hundred pounds apiece, and the two bulls locked in combat before us wouldn't even notice if they trampled a tent or its occupants in the scuffle.

There we stood, shivering, bare feet burning in the new frost, squinting in awe at the shadowy beasts clashing beside us. The bulls circled, assessing each other, and then lowered their heads and slammed. Antlers clacked as they locked heads, gigantic silhouettes straining, grunting, gouging divots from the earth each time they lunged. Hindquarters whirled by our tent as they quickstepped their ancient dance, oblivious to the world around them. In the end, we didn't get trampled, and our tent survived unscathed. But images from that September night fifteen years ago stay etched in my mind. I still remember steam rising from their breath in swirling clouds against the dark shadows. I even remember the smell, thick and musky, from the oil glands on the bulls' faces.

By any account, elk are magnificent beasts, icons of power and beauty. But most of what impresses us sticks out from the tops of their heads. It's the antlers that inspire our wonder. The weapons. The racks of elk, red deer, moose, and caribou have added regal splendor to the walls of royal halls for centuries. Indeed, no self-respecting chateau or castle would be complete without them. Antler-wielding stags are one of the most pervasive symbols in heraldic coats of arms, and mounted heads with antlers or horns grace the fireplaces of innumerable hunters' dens, sporting goods stores, restaurants, and bars, reflecting silent glory upon the ones who slayed them.

Obsession with animal weapons is nothing new. The earliest known paintings attributable to our species, inked onto the smoky walls of caves more than thirty thousand years ago, feature the branching antlers of stags, curved mastodon tusks, and rhinoceros and buffalo horns. Today, antlers and horns are embedded into the brand strategies of corporations ranging from single malt scotch (Glenfiddich, The Dalmore) and other spirits (Jägermeister, Moosehead Lager), to farm equipment (John Deere), firearms (Browning), automobiles (Porsche, Dodge), clothing (Abercrombie & Fitch), mountaineering gear (Mammut), sports franchises (Manitoba Moose, St. Louis Rams, Milwaukee Bucks, Texas Longhorns), and even pharmaceutical companies (Janssen) and investment firms (The Hartford, Merrill Lynch). No matter how you stack it, we love antlers and horns.

But why are antlers so impressive? What is it that captures our imagination and awe? It's not just that they're weapons—most animals have weapons of one sort or another. Tigers and lions have claws, eagles have talons, snakes have fangs, wasps have stingers, and even our household dogs have a respectable set of teeth. What strikes us about antlers is that they are *big*. The rack on a bull elk is forty pounds of bone erupting from the head in two curved beams, each adorned with as many as seven sharp tines. In the largest bulls, antlers tower four feet above the male and arch backward over half the length of the rest of his body. That's massive. And, although most of us never stop to think about it, we all know at some level that anything that big must also be expensive. In fact, the price bulls pay for their antlers is extraordinary, and they pay this price again and again since they shed and regrow their antlers anew each year.

Unlike the rest of the body, which takes years to grow to adult size, antlers in even the largest bulls go from nothing to full size in just a few months. Antlers grow faster than any other bone in any animal, and this record speed racks up record energetic costs. Estimates from antlers of a related species, fallow deer, show that while males are growing antlers they more than double their daily energetic needs. In addition, growing antlers suck up so much calcium and phosphorus—minerals that make up the bone—that the males cannot possibly get enough from their food. Instead, they leach these vital minerals out of other bones and shunt them to the antlers, depleting the rest of their skeleton so severely that they experience a seasonal form of osteoporosis.

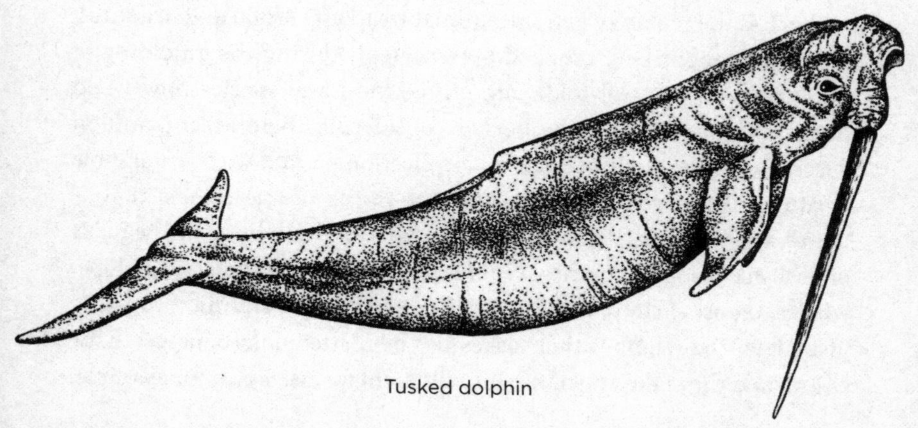

Tusked dolphin

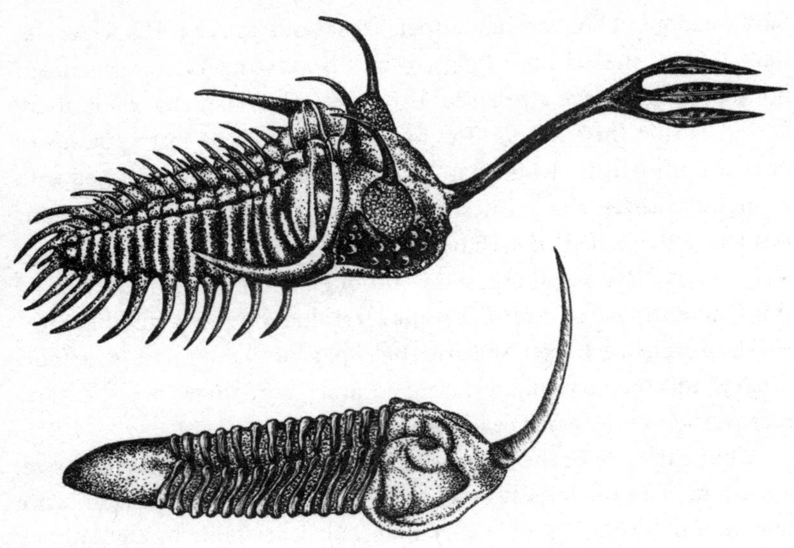

Trilobites with "horns"

Their bones get weak and brittle at precisely the time of the year—the rut—when they must hurl themselves against eight-hundred-pound rivals in incessant battles for access to females. By the end of the rut, the males will have fought so often and so hard that they've lost a quarter of their body weight, and they emerge from this season battered, starved, and brittle boned. If they cannot replenish their reserves in the few short weeks before winter, they'll starve.

Such is the reality of extreme animal weapons. Brutal and beautiful, extreme weapons have cropped up repeatedly during the unfolding of the history of life. All told, some three thousand species now wield them. That's a drop in the bucket considering there are 1.3 million described types of animal, but it's a collection packed with remarkable creatures. Early champions include the horns of triceratops, titanotheres, and trilobites, the tusks of mammoths and dolphins, and the racks of Irish elk. Today, extreme weapons are wielded by walruses, antelope, whales, crabs, shrimp, beetles, earwigs, plant bugs, and flies, to name just a few. The weapons themselves can be matted hair, bone, teeth, or chitin, and they take any number of different forms. Some, for example,

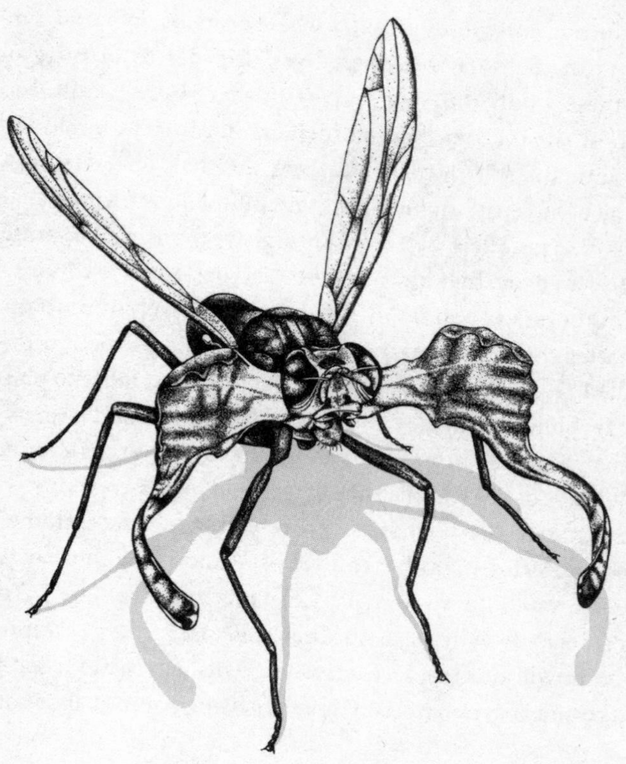

New Guinean moose fly with "antlers"

are overgrown versions of an existing structure, like a tooth or a leg. Others appear to have arisen de novo, as new bumps or knobs that became so large they formed their own distinct structure. In absolute size, they run the gamut from quarter-inch "antlers" on a New Guinean fly to sixteen-foot tusks on a mastodon. Yet, relative to the size of the individuals who carry them, all of these weapons are massive.

This is a book about extreme weapons, structures so gargantuan and bizarre they look like they shouldn't be possible—so awkward that the animals who bear them ought to tip over, or trip, or get tangled each time they attempt to move. Why are these weapons so big? What happens to animals once their weapons get this big? And is there such a thing as *too* big? To answer these questions, we'll delve into the murky forests and mountainsides where animals do battle, immersing ourselves in the details of their lives in order to identify patterns: things these wildly different species all share in common, and things that illuminate the logic behind such extraordinary animal forms.

We humans are animals, too, and no book on extreme weapons would be complete without an examination of our own arsenals. We'll see that the parallels between animal weapons and manufactured weapons run deep. In both cases, the vast majority of weapons are relatively unimpressive and small. But, here and there, circumstances arise that shatter the norm, sparking bursts of rapid escalation in weapon size called "arms races." Very specific factors must fall into place before weapon evolution launches into one of these races, and it turns out that the same special circumstances triggering arms races in animals also prompt humans to manufacture bigger and bigger weapons.

Once started, both sorts of races quickly lead to extreme weapon forms—staggering in both size and cost—and along the way they each progress through the same sequence of recognizable stages. Analogous circumstances even bring about their collapse, as huge weapons come crashing down and the race dissolves. Ultimately, we'll see that our animal counterparts can teach us a surprising amount about ourselves.

• • •

Animal Weapons is necessarily a book about evolution: the gradual turnover within populations that leads to changes in animal form over time. At its most basic, this evolutionary process is a simple one. Entities exist in multiple forms—they vary from individual to individual—and their individual characteristics are transmitted from one generation to the next. The transmission of this information is efficient but it's not perfect, and mistakes get made along the way. Errors in the transmission of information lead to new characteristics—novel variations that crop up in the population from time to time. New variants now exist side by side with earlier forms, competing with them for resources and reproduction, and only some of them will persist.

Whenever individuals differ in how successful they are at propagating their kind, evolution occurs. This can be due to chance—random deviations in the reproductive success of types—or it can result from selection; individuals with some characteristics perform better than individuals with other characteristics, and as a result they leave behind more copies of their kind. Such a process, iterated over generations, results in the eventual replacement of inefficient with more efficient forms. The population evolves as the least effective types are culled and replaced by more effective types. As transmission errors yield new types of individuals, they inject fresh forms into the mix. If the new individuals do worse than the rest, they gradually disappear. If they do better, then the new forms can spread, replacing earlier forms in the process. This turnover is evolution.

Huge animal weapons look too awkward to be favored by selection, and in the vast majority of cases, this appearance isn't misleading. Big weapons *are* ungainly, and most individuals fare poorly with them. For most weapons in most species, selection favors modest size and minimal cost—enough tooth to bite or grab prey, for example, but not enough to slow you down or impair your ability to maneuver. What this actually means is that selection on weapons is balanced: bigger weapons may be better for stabbing or biting, but they are also more expensive to produce and more difficult to carry. The result is a tenuous balance,

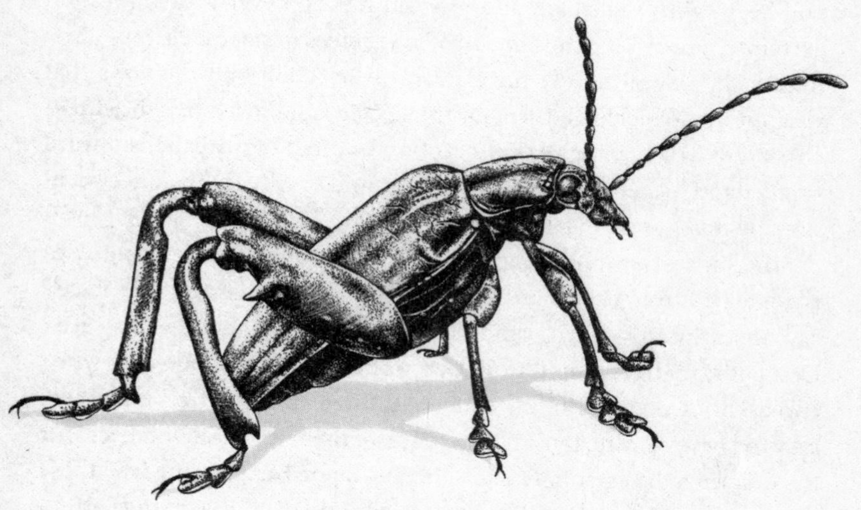

Frog-legged leaf beetle

much like a game of tug-of-war, which accounts for the multitude of unimpressive weapons adorning the vast majority of animal species.

Yet, every so often this balance is tipped. Here and there, sprinkled across the tree of life, are animal lineages where modesty of proportion was cast aside. The evolution of weapons in these species surged forward unencumbered. Gone were the shackles of balancing selection, and all that remained was selection favoring bigger and bigger weapon sizes. Here, individuals with the most grotesque or extreme armaments beat opponents with smaller weapons. In so doing, they secured opportunities to breed. Their progeny, as impressive in their weapons as their parents, quickly replaced earlier forms and advanced the population another notch in weapon size. As soon as the next new innovation arose—an even larger or more complex variant of the existing design—the process repeated itself. Again and again, individuals with the newest and biggest weapons won, replacing those with earlier and smaller forms, and this process pushed the population still further on the path to extreme. In short, the weapons of these species got caught up in an arms race.

Weapons swept into arms races begin small, but over time they get bigger and bigger, increasing in size faster and faster. So, too, this book will begin with the small, and proceed through ever-greater weapon extremes, assembling the biology of arms races in stages. Part 1, "Starting Small," examines the mechanism of selection and the ways that weapon designs change through time. This is followed by a brief discussion of balance, the reason selection so often tugs weapons toward smaller and bigger sizes simultaneously, and of why this balance occasionally tips.

The biggest animal weapons result from males competing with rival males over access to females. Part 2, "Triggering the Race," explores why this is true, and how competition leads to arms races. Competition drives rapid evolution of huge weapons, but only when two additional conditions are met. All three "ingredients" are necessary to trigger a race. By exploring each of them in detail, I'll reveal for the first time the essential logic of huge weapons, explaining why these structures evolve in some species but not others, and why they attain such incredible proportions.

Part 3, "Running Its Course," explores what happens after an arms race is triggered, describing in detail the predictable sequence of stages characteristic to the evolution of nature's biggest weapons. Staggering costs, deterrence, and cheating all appear as intuitive milestones along the way, and costs and cheaters erode payoffs to huge weapons in ways that can set the stage for collapse. Appreciating how arms races unfold reveals a great deal about the weapons themselves, about the conflicts and contests surrounding their use, and about what happens when weapon evolution goes too far.

Finally, Part 4, "Parallels," tackles head-on the evolution of extreme weapons in our own checkered past. Although I am a biologist (not a military historian), and this book is primarily concerned with diversity and extravagance in animal weapons, the parallels with our own armaments are too striking and exciting to ignore. Every element of arms races, from the conditions critical to launching a race, to the sequence of stages unfolding along the way, lends itself to comparisons between animals and people. Manufactured weapons are not inherited in the strict sense—instructions for their production are not encoded in DNA—and they are assembled in factories rather than wombs. The competitions they resolve are less likely to yield opportunities to mate, and success is often measured in currencies other than increased numbers of offspring. Nevertheless, human weapons change in shape, capability, and size over time, and the directions of this change are sculpted by forces of selection astonishingly similar to the forces shaping weapon evolution in animals. Arms races are arms races, and the natural history of extreme weapons, it turns out, is precisely the same.

PART I

STARTING SMALL

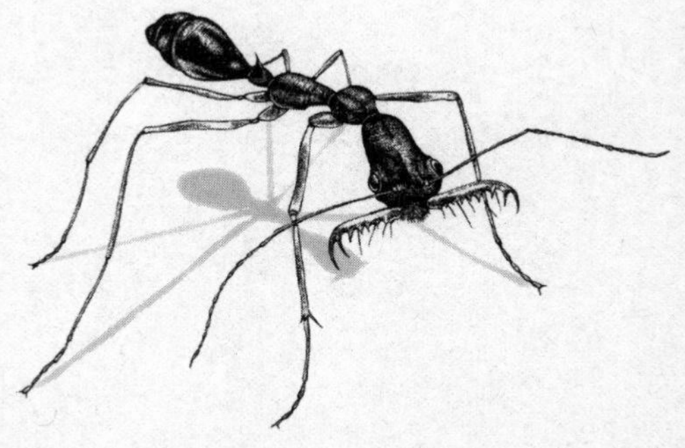

1. Camouflage and Armor

It is November 1969, and dark. Moonlight glints off of tree branches, casting small shadows beside pebbles and twigs on the fresh dirt. A tiny metal door opens and two mice rush out, like gladiators released into the Roman Colosseum. They dash into the darkness in search of cover, but there isn't any, and only one of the mice will survive this race. Above them an owl sits on a perch, watching. Its head snaps into position as it spots a mouse and, with a graceful swoop utterly devoid of sound, it attacks. One moment both mice are running. The next, just like that, one of them is gone. Droplets of blood on the dirt are the only evidence of what has occurred.

Six concrete enclosures lie side by side, encased in chicken wire to keep the owls from escaping. Twelve feet wide and thirty feet deep, they are more vast to a mouse than Mile High Stadium is to you or me. In three of them the soil is rich and dark, imported from a nearby field. In the others, the soil is sandy and pale, trucked from coastal dunes. Otherwise, the enclosures are the same, and each houses an owl, patiently waiting. Over and over the race is repeated, as pairs of mice—one brown, one white—sprint across the dirt. All told, almost six hundred mice will rush into the South Carolina night, all to answer one question: which color mouse will the owls catch first?

Owls eat astonishing numbers of mice. When owls feed, they pack fur and bones and other indigestible parts of their prey into their gizzards, coughing them up later as dense little pellets that they spit onto the ground. Diligent biologists can harvest these pellets and pick through them, counting and identifying bones, to reconstruct an owl's diet on any given night. A single owl can eat four or five mice a night, and well over a thousand in a year.[1] Scaled up to the landscape, owls kill between 10 and 20 percent of mouse populations in a typical year—up to one-fifth of all mice die in the talons of an owl.[2]

Despite the brutal toll exacted by owls and other predators, oldfield mice thrive across the southeastern United States. They live in abandoned corn and cotton fields, along hedgerows, in forest clearings, and throughout all sorts of shrubby fields. These mice also live along coastal beaches in sand dunes tufted with coarse grasses, and they have colonized many of the small offshore barrier islands of Alabama and northern Florida.

In the mid-1920s, Francis Bertody Sumner, the leading mouse biologist of his day, heard about the strange white mice of Florida beaches. He worked his way across their range, meticulously sampling animals from population after population. Some he brought into his lab to breed, but most he killed, stretching their little pelts for archiving in museum collections. The pattern he documented was striking: mice from inland populations—stubble fields and clearings across Alabama, Tennessee, South Carolina, Mississippi, Georgia, and interior Florida—were dark brown in color, much like that of other field mouse species found elsewhere in the United States. But along the coasts and out on the sandy offshore islands, the mice were white. And, if you marched a line from inland to coast, there was an abrupt transition separating brown mice from white. The boundary fell about forty miles inland from the shore, and it tracked the coast like a contour line on a map.[3]

Sumner noticed that around this transition zone, the soil also changed color. Inland, the soil was loamy and dark, filled with organic detritus from decaying vegetation. Near the shore, soils were sandy and white—in some cases mice lived on dunes of bleached sand so bright they resembled giant mounds of sugar. Ninety years later, Lynne Mullen and Hopi Hoekstra, biologists at Harvard University, retraced Sumner's steps and sampled the populations again. A thousand mouse generations separated the two samples, but the pattern held. Soil color changed abruptly from brown to white, and mice matched this transition with a shift in fur color.[4] Brown mice lived inland, and white mice lived on the beaches.

In all other respects, inland and beach mice are similar. They make the same kind of burrows, for example. They cut in at an angle, leveling off into a horizontal nest chamber about a foot below ground. Many of them also make an "escape hatch"—a vertical tube that extends from the nest chamber straight up, stopping just an inch below the surface.[5] If a snake or weasel pokes into their burrow entrance, they can "explode" through the thin soil capping this shaft to escape. Inland and beach mice eat the same foods, including insects, seeds, and the occasional berry or spider. By all metrics except color, these mice are the same. So why are the coastal mice white and inland mice brown?

This was the question Donald Kaufman sought to answer with his gladiatoresque doctoral dissertation experiment that November back in 1969. Over and over, night after night, he released dark mice and white mice into cages side by side. Each time the owl snatched one of the mice, Kaufman recorded which one died, and which survived. He showed that both soil color and mouse color mattered. When the mice dashed across dark soil, the white mouse was most often taken. When the soil was pale, the pattern was reversed. Owls snatched the darker mouse. There were additional nuances to the owls' behavior. For example, on the darkest nights the pale mice fared especially poorly on the dark soil. Their white fur contrasted starkly with the blackness of their surroundings. On the other hand, bright, moonlit nights and light soils made the dark mice stand out most sharply. Mouse survival depended to some extent on ambient moonlight and local conditions but, overall, the pattern was clear: mice whose fur color contrasted with their backgrounds got eaten.[6]

Hopi Hoekstra and her colleagues completed this story by tracking down the genes, and even the particular mutations to these genes, responsible for fur color in mice.[7] Once Hoekstra's team knew the molecular machinery responsible for genetic variation in mouse color, they could reconstruct precisely how mice evolved in response to recent changes in the direction of natural selection. Most oldfield mice are brown, and this color is favored by selection across the majority of fields inhabited by this species. At some point in the past—possibly as recently as a few thousand years ago—mice spread into open areas along both the Gulf and Atlantic coasts, where they dug their burrows into sand dunes and grassy embankments. Beach mice now raced across a vastly different background than their inland ancestors, and in these new environs dark mice got plucked from the sand.

By chance, some of the beach mice carried in their DNA new mutations to one or both of two genes involved in the production of dark pigments. Mice inheriting these mutations carried copies of the pigment-influencing genes that were just a little bit different from the copies carried by other mice (alternative versions of a gene are known as alleles), and as a result they developed with lighter fur. Mice bearing the new alleles survived better than mice inheriting the ancestral versions of the genes, and these survivors populated the beaches with their pups. Over time, mice with the new alleles increased in frequency, while those with the original alleles disappeared, and the result was an evolutionary shift from dark to white.

• • •

Camouflage might seem an odd place to begin this book. But weapons come in many forms, and not all of them function offensively. A U.S. Army infantry soldier marching into conflict carries all sorts of gadgets that contribute to his or her efficacy in battle. Not counting specialized weapons such as grenade launchers or squad automatic weapons (SAWs), the primary weapon is an M4 carbine assault rifle with removable bayonet.[8] Along with this, soldiers carry fragmentation grenades, knives, food, water, and first aid kits. They wear body armor (vests with plates of finely woven Kevlar designed to protect against bullets and heat), a helmet, and cloth uniforms—"camo"—with color patterns designed to blend with the surrounding landscape.

Many of these items function for defense, rather than offense, but they are no less critical to troop success in combat; for this reason they can be considered weapons. Although this book is principally about extreme weapons—the biggest tools in nature's arsenals—it begins with these other weapon types, including animal analogs to camo and armor, which we'll cover in this chapter, and to lightweight and portable small arms, which we'll get to in chapter 2. Each of these animal examples has been studied unusually thoroughly, providing clear insights into the processes of selection and evolution. All of them also have intuitive parallels with man-made manufactured weapons.

Obviously, blending with backgrounds is essential for soldier survival for precisely the same reasons that it is in mice (imagine conducting a night operation wearing white winter camo). In fact, in 2003 the U.S. Army used a process not unlike Kaufman's experiment with owls to determine the most effective camouflage patterns for our troops. More than a dozen color and pattern types were assessed against urban, desert, and woodland environments, to identify uniforms least likely to stand out.[9] Some of these tests were conducted at night, where they showed—just like Kaufmann—that being too dark on moonlit nights could be deadly. Modern enemy soldiers, it turns out, are a lot like owls. They have phenomenal nighttime vision thanks to the spread of night-vision goggles and other technologies. As a result, black has been eliminated from most camouflage patterns.

Ideally, the uniform selection process should have unfolded just like owls selecting for mismatched mice, with the population—in this case, the army—evolving toward the best camouflage possible. Unfortunately, politics and the economics of mass production intervened. Rather than choose several different types of uniforms, each the best available for a particular habitat, the army opted for a single Universal Camouflage Pattern (UCP).[10]

This may have solved logistical problems of production and distribution, but it also caused our troops to sometimes stand out when they were supposed to be blending in. After all, the solution with mice was two colors, not one, and the reality of diverse combat habitats is that no one pattern blends well in all places.

It didn't take long for our troops to complain,[11] and by 2009 it was obvious to everyone that the UCP was performing terribly in

Afghanistan.[12] The army then rushed to develop a new pattern, called "Operation Enduring Freedom Camouflage Pattern" (OCP) for soldiers deployed in Afghanistan, which it began issuing in 2010.[13] Special Forces soldiers, incidentally, are not subject to the same constraints of mass production, and these units have diverse and effective uniforms to choose from, depending on the mission. Military units in other countries also base pattern choices on advanced tests of detectability.[14]

The brutal reality of life and death on the battlefield has provided a sort of natural selection for military uniforms. Many versions are tried, some perform better than others, and patterns performing the best are (usually) selected for further use. Despite various hiccups along the way, few would disagree that modern uniforms are vastly improved over those worn in earlier wars. WWII uniforms were better than those of WWI, and uniforms today are better than those used in Korea or Vietnam.

• • •

From lizards and desert beetles that look like little pebbles, to giant tropical katydids resembling decaying leaves, camouflaged animals are evolutionary legacies of the same basic process: natural selection by visually searching predators. Predators don't just drive their prey to match colors with their backgrounds; they drive them to behave in new ways. When and how an organism moves can influence how vulnerable it is to predators. Animals that panic, dashing from their hiding places at the wrong time, or animals that walk or fly with the wrong gait, can break camouflage with deadly consequences. A leaf-mimicking katydid would stand out if it whirred through a forest clearing in broad daylight. In fact, these katydids forage at night. During the daytime, they rest on branches nestled amid similar-looking leaves. If they do need to move in the daytime, they lurch with a swaying gait just like the back-and-forth flutter of a leaf in the breeze. Place one of these animals on an exposed, flat table, and this gait looks absurd. Place it on the branch of a bush and the animal vanishes; the katydid's movements, combined with its shape and color, make it almost indistinguishable from the surrounding leaves.

Resembling a leaf is a relatively passive defense—hardly a "weapon"

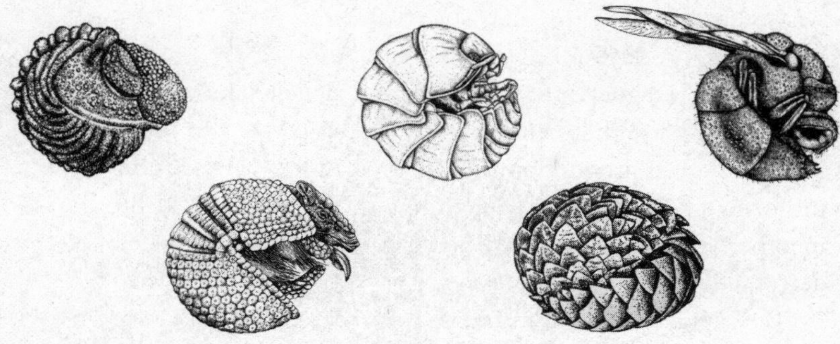

"Curling up" defensive postures evolved in tandem with armor in trilobites, pill bugs, cuckoo bees, armadillos, and pangolins.

proper—as is mouse fur blending with soil. Other animal defenses are much more formidable. Many animals use chemical weapons against their predators, either synthesizing toxins, or extracting (and sometimes modifying) toxins from their food.[15] Some caterpillars ooze droplets of poison from glands near the bases of tiny barbed, needlelike hairs. Poison-dart frogs pack toxins into their skin, foam grasshoppers disgorge foul-tasting bubbles from their armpits, and bombardier beetles spray jets of exploding acid from their anuses.

Still other animals use armor to protect themselves. Like the jointed metal breastplates and shields sported by Roman centurions and medieval knights, many animals cover their bodies with tough plates of compacted hair, bone, or chitin (the main ingredient of insect and crab exoskeletons). Turtles and crabs may be the most familiar examples, but armor plates also protect armadillos, pangolins, pill bugs, and tortoise beetles, and similar shells protected the extinct glyptodonts and ankylosaurs as well.

My personal favorite among the wild array of defensive weapons are spikes and spines that jut from the flanks and backs of prey—blades of bone or chitin sharp enough to puncture the mouths of predators and tear delicate linings of digestive tracts. Daggerlike spines protect all sorts of animals, ranging from porcupines and hedgehogs to spiny crabs, long-spine porcupine fish, and katydids.

Three-spined sticklebacks swim in shallow waters along the coasts

Porcupine spines are effective defensive weapons.

of Europe and North America. These finger-sized fish rely on both sharp spines and armor to protect them from predators. Rigid spines project along their back and from their pelvis, and a row of bony plates adorns their flanks. Here, as with oldfield mice, biologists understand both the genetic basis for variation in defensive traits—the genes responsible for heritable variation in spine length and plate size or number[16]—and they understand how this variation has contributed to rapid weapon evolution in the face of natural selection. This time, however, the traits are rigid bony outgrowths rather than pigments altering the color of fur. Understanding how these weapons evolve sets the stage for the much larger outgrowths we'll consider in later chapters.

As with all evolutionary tales, the stickleback story begins with variation. Some sticklebacks invest more in defensive weaponry than others, resulting in fish-to-fish differences in the length of pelvic spines and in the size and number of body-armor plates. Not surprisingly, this variation in weapon size influences fish survival. Long spines make sticklebacks difficult to swallow (think splintered chicken bones lodged in a dog's throat), and armor plates protect sticklebacks whenever predatory fish make the mistake of trying to bite them. Almost 90 percent of attacks on sticklebacks fail. But before spitting them out, predators chew sticklebacks rather harshly. A stickleback's armor plates act like shields, reducing the extent of injuries from these bites.[17]

While most sticklebacks live in the ocean where predators are common, some inhabit freshwater lakes, and here their evolutionary story is different. Ocean levels fluctuate greatly over time, and during periods of high water fish spill into inland reservoirs, where they end up trapped as the water recedes. Inland fish experience very different patterns of selection from their marine ancestors, and in lake after lake, their weapons have changed as populations adapted to their new locales.

Fossils provide a road map of this weapon evolution. In fact, so many stickleback fossils have been preserved that they provide an almost unparalleled paleontological record of change in weapon size through time, as layer upon layer of fish corpses piled into the mud at the bottoms of lakes. Michael Bell, a biologist at Stony Brook University, studies this temporal progression of fish in a Nevada lake bed, where he and his colleagues reconstructed approximately one hundred thousand years of stickleback evolution in 250-year slices.[18]

In the beginning (well, the first eighty thousand years of their one-hundred-thousand-year window), Nevada lake sticklebacks had almost no protective weapons (only one dorsal spine, rudimentary pelvic spines, and very few lateral plates). But then, eighty-four thousand years into the time sequence, this type of stickleback was replaced entirely by armored sticklebacks, meaning three long dorsal spines and full pelvic spines. Bell suspects that marine fish flooded into the lake around this time, because both forms co-occurred for about one hundred years before the early fish type disappeared. Remarkably, over the following thirteen thousand years, the defensive structures in this new fish regressed: in graded steps through time, the spines got shorter and shorter, until by the end of this period the new sticklebacks resembled the earlier form that they'd replaced. Lake-bound fish lost their weapons.

Freshwater and marine sticklebacks differ in the length of their spines and the number of bony armor plates.

Today, sticklebacks in many lakes lack defensive weapons. Dolph Schluter and his students at the University of British Columbia found that lake habitats have far fewer predators than marine habitats, and this appears to relax the pattern of natural selection for larger plates and longer spines.[19] With fewer predators, lake fish benefit less from large weapons than marine fish. Armor also costs more in lakes than it does in the ocean. Low freshwater concentrations of the ions necessary for bone growth mean that fish pay a higher price for mineralizing bony plates in lakes. Unarmed sticklebacks are larger as juveniles and begin breeding sooner than their armed counterparts. In freshwater, it appears, the costs of long spines and large plates are steeper than the benefits they provide.

Of course, every story has exceptions. But with sticklebacks, the exception proves the rule. Dan Bolnick has been studying sticklebacks in Lake Washington, where fish have much bigger weapons than in other lakes. Bolnick found that this shift in armor happened very recently—fish samples collected prior to the 1960s had reduced armor typical of other lake sticklebacks. Efforts to stem pollution in this lake resulted in a dramatic improvement in the transparency of the water, and in these especially clear waters, introduced trout began to feed in earnest on sticklebacks. More predators translated almost immediately into bigger weapons.[20]

• • •

Since the beginnings of recorded history, soldiers have sought protection from enemy weapons. Their armor mirrors the armor of stickleback fish and other animals, evolving for similar reasons and in surprisingly similar directions.

The earliest forms of body armor were the shield—made originally from animal hide and later from leather stretched over wood—and protective clothing made from leather, padded fabric, wicker, or wood.[21] Over time, as the technology of weapons changed, so did the shapes and styles of protective body armor. The first manufactured weapons were sharp, fire-hardened pointed sticks and spears tipped

Roman legionary wearing a leather cuirass with attached metal plates.

with flint blades. Knapped flint could slice flesh, but blades shattered easily—stiffened leather provided adequate protection against flint for thousands of years.[22] Metallurgy brought with it tougher weapons—first bronze, which was soft and dulled quickly, and later iron—and against these weapons, leather armor was not as effective. Armorers began attaching metal rings or scales to the outsides of leather garments to block against the stabbing or slicing blows of metal-tipped pikes and swords. Ancient Greek soldiers wore a leather cuirass covered with front and back plates of hammered bronze, and Roman legionaries wore leather cuirasses stitched with metal plates arranged in overlapping tiers much like the scales of a fish (these warriors also wore helmets and carried bronze shields).[23]

By the time of the Crusades (1100–1300 CE), protective shirts of mail—intricately linked chains of iron rings—had become the standard attire for battle in Europe. Mail could block penetration from most strikes by metal blades, but shock from the impact was still severe. Soldiers often wore thick clothing of padded fabric or leather underneath mail. On top, they added scaled metal and leather cuirasses and helmets. Soon, plates of iron were added to vulnerable areas, such as the elbows, shoulders, and legs. By the end of the fourteenth century, full suits of plate armor had replaced mail—think "knights in shining armor."[24] Plate armor predominated until the sixteenth century, when its use in warfare was rendered obsolete by gunpowder and firearms.[25]

From the beginning, the evolution of protective manufactured armor was shaped by a balance between benefits and costs. As weapons became more dangerous, the thickness and toughness of armor increased, but so did its bulk and weight. On the one hand, armor could protect a soldier, but on the other it restricted his movements and slowed him down. A suit of chain mail weighed up to fifty pounds, not counting the heavy leather beneath it. A helmet alone could weigh twenty pounds, and helmets were so hot and suffocating that knights generally carried them on the pommel of their saddles until the last minutes before battle.[26] Plate armor was a huge burden; getting knocked over or unhorsed could mean death for a knight, since he couldn't get back up without assistance.[27] By the end of the sixteenth century, crossbows and longbows had already called the efficacy of armor into question (arrows striking straight-on could penetrate), and the spread of gunpowder sealed its fate.[28] As with the bony plates of lake-bound sticklebacks, once the benefits disappeared, armor was no longer worth the price. Although plate armor thick enough to stop a bullet was possible, nobody could wear it because the plates were too ungainly and heavy.[29] So body armor all but disappeared from the battlefield for four hundred years, until a new invention: Kevlar.[30]

• • •

In armor we see all of the processes that matter for the evolution of extreme weapons: individuals vary in the extent of their armament; these differences in weapon size affect the performance of their bearers (survival, growth, and reproduction in sticklebacks, and survival in soldiers); and, as a result, the sizes and shapes of these weapons evolve rapidly and dramatically over time. Weapons such as armor come at a price, and sometimes, when this cost is high enough, individuals with small weapons fare better than those with big ones. Indeed, most of the time, for most weapons, bigger is not better.

2. Teeth and Claws

Mountain lions are still common where I live in Montana, and for many Montanans these cats are part of why we choose to live here. There's a tiny rush you feel every time you step out into the wilderness, a reminder that in some places we're actually not the top of the food chain. I had my first face-to-face encounter with a mountain lion this past December, hiking on the ridge behind my house. I'd known they were there, having seen paw prints in fresh snow on countless winter mornings and from stumbling upon buried kills in the forest nearby (lions cover a carcass with pine needles and branches so they can come back to it later). I also knew they were there because I was actively looking for them; a few years before, I'd placed a motion-activated trail camera next to a little spring in a ravine behind my house. Each week I hike up the mountain to the spring to swap out the memory stick and then sort through the thousands of pictures of magpies, deer, skunks, bears, eagles, and, of course, mountain lions.

That morning last December, I was coming over the hill and starting down to the spring when my dog tore ahead of me. The cat he was chasing leapt to the nearest pine, shot up into thick branches, and disappeared. For a stunned moment I was impressed with my dog—a flat-coated retriever; a family dog, not actually a mountain dog—and then it was time to assess. No bear spray that day, no camera, no knife, not even a leash for the dog. Basically, I was utterly unprepared, and the cat looked small enough that I knew it might not be alone. A mother lion could spring from the bushes behind me and, unless I saw it coming, I wouldn't stand a chance. So I took my dog by the collar and we back-tracked over the hill and down to my house. When, a half hour later, I hiked back up better prepared, I found no trace of the cat whatsoever. But when I retrieved the memory stick from the camera and opened the files, there were pictures of *two* lions that morning, not one. Big cats rarely choose to hunt humans, but they can be devastatingly efficient when they do. I was lucky I'd retreated.

Silent, fast, and deadly, cats are quintessential mammalian predators, yet their weapons are relatively small. To appreciate why, consider what it is these animals do. Canada lynx, for example, stalk alone through vast boreal forests, soundlessly sweeping back and forth across the snow in search of their preferred prey, snowshoe hares. Overtaking a hare is no easy feat. Finding them is difficult because their fur blends seamlessly with their backgrounds. Hares molt from brown to white during the winter when their landscape becomes snow covered—their solution to the problem of camouflage on more than one background.

Once found, hares have to be caught. Massive hind feet give them a tremendous advantage, including an almost unmatched potential for acceleration. Galloping hares can top forty-five miles per hour, making them second only to pronghorn antelope as the fastest land mammals in North America. On top of that, the power in their long hind legs lets hares change directions erratically, without sacrificing acceleration or speed.[1] There's a reason it's the hare in Aesop's fable "The Tortoise and the Hare."

Given the hare's speed and agility, it's no surprise that lynx very often fail to catch them. Footprints in fresh snow tell the stories of these encounters, revealing where each hare was flushed, how far it ran, and who won the race. In one study following hundreds of miles of lynx tracks over five years, the cats caught their prey in only one out of every four chases; in a similar study lynx captured a hare only once every four or five days—barely enough to sustain their body mass.[2] Even in good years most lynx hunts end in vain. In bad years, lynx fare much worse.

Snowshoe hare populations fluctuate wildly in number, with over forty times more animals in "boom" years than in "bust" ones. These population cycles mean that every eight to ten years lynx experience drastic shortages in their food supply and, in these lean years, starvation is rampant. Kitten survival plummets from 75 percent in years with abundant hares to 0 percent when hares are scarce.[3] The difficulty of catching prey and the periodic shortages in abundance of prey together result in intense selection for improvements in the lynx's hunting performance, placing a premium on the weapons necessary for successful hunts.

In fact, much of the diversity of predators can be understood from an examination of their weapons and the ways in which these structures have adapted to different types of habitats and prey. The history of the mammalian order Carnivora, for example, is a story filled with successes and failures all defined by the evolution of their weapons: forelimbs, claws, jaws, and, especially, teeth.

• • •

The earliest meat-eating mammals appeared not long after the disappearance of the dinosaurs, roughly sixty-three million years ago. These carnivores were scrawny semi-predators who ate a mixed diet, and their teeth were correspondingly all-purpose. *Vulpavus*, for example, was a small, ferret-sized animal with a lanky body and slender tail that probably fed on insects, spiders, lizards, birds, and small mammals such as shrews.[4] The principal instruments in this early carnivore's dental toolkit were canines, incisors, and a row of premolars and molars along each jaw. Already, by the appearance of the first carnivorous fossils, subsets of teeth had begun to take on specialized functions. Canines were longer than the other teeth, and effective at apprehending and killing prey. The pointed premolars could grip and hold prey, and the molars could slice and crush a carcass during feeding.

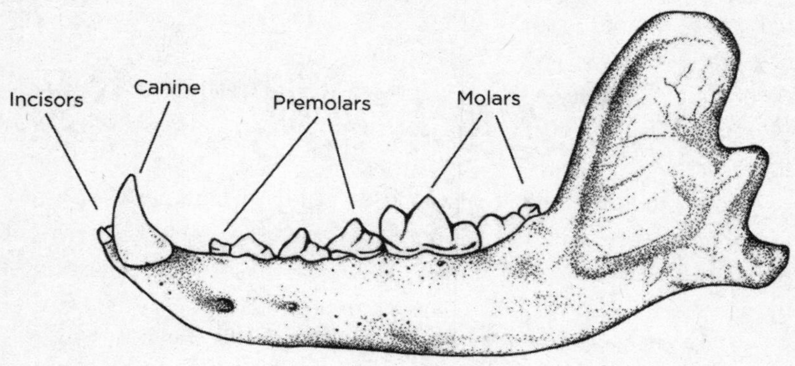

Subsets of carnivore teeth became increasingly specialized for specific tasks, such as piercing, slicing, or crushing.

Over time, these subsets of teeth evolved to become increasingly efficient for their specific tasks. At the same time, the nature of these tasks began to change as the number of carnivore species multiplied. Many species began to specialize on narrower and narrower subsets of prey, and the demands on their teeth started to differ from species to species. Carnivore teeth evolved in new and different directions, depending on the particular diet and hunting habits of each species. Although some species retained relatively basic tooth shapes suitable for omnivorous diets, many, including wolves, hyenas, cats, and sabertooths, diversified into efficient "hyper-carnivore" predators specialized for diets that consisted of meat only.[5]

Wolves are the "jacks-of-all-trades" of the hyper-carnivores. Their long, slender jaws snap shut with amazing speed, and their sturdy canines grip flanks or legs of large prey as they wrestle them to the ground. Wolves hunt in packs and, by pulling from several directions at once, they can topple animals far larger than themselves. After the kill, wolves tear into carcasses using dual-purpose molars. Sharp outer edges work like shears to slice through sinew and flesh, but these teeth are still broad enough to crush small bones.[6]

Hyenas also hunt in packs, but their jaws are very different from those of wolves. Hyena canines are relatively short, and their molars have lost the "dual-purpose" functionality of their ancestors. The slicing edges of hyena molars are gone. Hyenas are bone crushers who feed

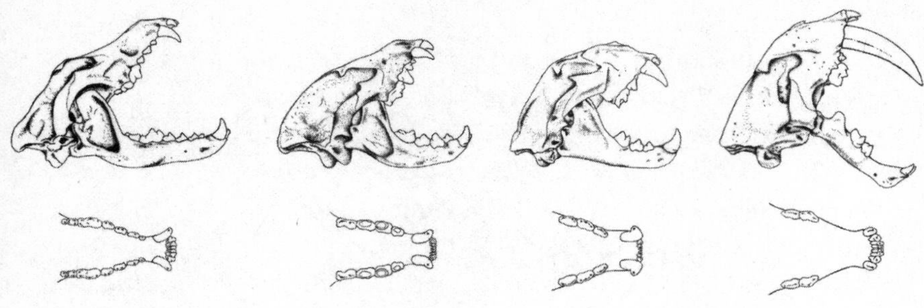

Wolves, hyenas, cats, and sabertooths differ in the relative sizes and shapes of their teeth.

primarily on marrow, and their teeth are wide and strong with rounded, dome-shaped caps. Their faces and jaws are squat, giving their teeth a huge mechanical advantage. It's basic physics: the nearer an applied force is to the joint of a lever, the stronger it will be. Teeth on short jaws aren't anchored very far from their hinges, resulting in slow speeds but powerful closing forces (contrast this with wolves, where canines positioned at the ends of long jaws snap shut faster, but without as much force). In hyenas, jaw-closing speed appears to have been traded for increased closing force. They have a tremendously powerful bite, and because of the shapes of their teeth, they use this bite for cracking bone rather than puncturing or slicing flesh.[7]

Cats also have snouts and jaws that are relatively short, favoring the mechanical power of jaw closure over speed. And, like hyenas, their molars have become specialized for just a single task. But this task is slicing, rather than crushing. Cat molars are narrow and sharp—useless for breaking limbs or bones, but ideal for cutting through flesh. And, while the primary weapons of hyenas are their molars, in cats the weapons are canines. Cats use their canines to puncture thick hides and sever the spinal cords of prey.[8]

Cats are specialized in another way. They can supinate their forelimbs—twist their wrists and pull the pads of their feet inward, facing their bodies. Flexible forelimbs let these animals cling to their prey and position themselves for the careful execution of their powerful, killing bite. Their canines are long and narrow, which makes them great for piercing but vulnerable to breaking if they get yanked to the side. A cat that can hang on to its prey in the throes of an attack can pierce its long teeth straight through the skin in a calculated stab. Failure to hold on results in torque during the bite, and may cause the canines to snap.[9]

Thanks to their flexible forelimbs, cats are unusually agile, able to pounce and, like the lion behind my house, climb trees.[10] (The old adage that a cat will always land on its feet is more apt than most people realize.) But as deadly as cats may be, they pale beside their extinct relatives, the sabertooths. The canines of sabertooths were truly massive: ten-inch daggers that could sever the spine of a mammoth. Sabertooth teeth wouldn't work without severe adjustments to jaw and skull shape, and to body posture. Over time, upper jaws became much shorter—even shorter than in other cats—producing powerful bite forces by bringing the canines ever closer to the hinge. Sabertooth jaws were thick, and the hinges could swing open to an unusually wide gape. Sabertooths had to pull their lower jaws all the way out of the way, as if releasing the bottom plate on a stapler, before they could sink their big teeth into prey. Finally, the shortened face and compressed skull tipped the head back, so the canines pointed forward during an attack.[11] All of these modifications rendered these carnivores among the most deadly ever to live, but the postural and head-shape changes came at a steep cost. They made running—and basically all movement—cumbersome and awkward.

Extreme tooth size allowed sabertooths to kill larger and larger prey. In a time when titanotheres, giant sloths, and mastodons abounded, this advantage would have been considerable. Saber-toothed predator forms arose in at least four mammalian lineages, first within two now-extinct carnivore groups, the creodonts (*Apataelurus* sp.) and nimravids (*Barbourofelis fricki*), then within the cats (for example, the scimitar-toothed and dirk-toothed cats), and finally, within the marsupials (*Thylacosmilus atrox*). We associate most living marsupials with Australia, but these pouch-bearing mammals once ranged over much of the world, and *Thylacosmilus* actually lived in South America.

Well-preserved specimens of the dirk-toothed cat *Smilodon fatalis* from the La Brea tar pits indicate that it was smaller than a modern lion, but more than twice as heavy (six hundred pounds), with a bobbed tail.[12] These short, stout animals could never have chased down prey and

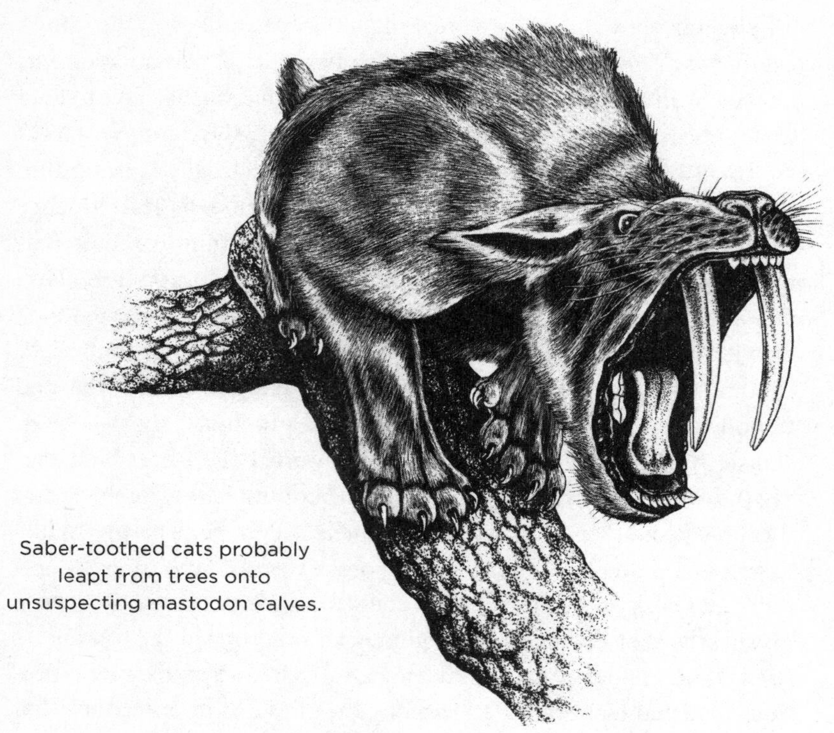

Saber-toothed cats probably leapt from trees onto unsuspecting mastodon calves.

almost certainly ambushed them from close range. Fossil bone beds suggest that sabertooths specialized on lumbering prey such as camels and young mammoths and mastodons, and the shapes of their forelimbs suggest they leapt onto the backs of these behemoths from trees.

• • •

Carnivore teeth aren't small because they couldn't or didn't evolve. They're small because individuals with unusually large teeth performed poorly when hunting their particular prey. Teeth and other major structures are almost always subject to trade-offs—a balance of opposing forces of selection. Bigger weapons may be better for killing prey, but they may also prevent an animal from catching prey in the first place. Individuals with unusually large weapons surely crop up from time to time in predator populations, but their hindered performance at critical tasks such as overtaking prey causes them to fare poorly and, over time, these extreme weapon forms are likely to disappear.

Sabertooths are a case in point. In each instance where canine evolution proceeded to this extreme, the enlargements to these teeth required dramatic adjustments to jaw and skull shape. Opening the jaws wide enough wasn't possible without modifications to the hinges, and sinking teeth this long into the neck or throat of prey required a severe backward tilt of the head.[13] Sabertooths could not run fast. They were simply too awkward. Such big weapons could never have arisen in carnivores who relied on speed to hunt down prey.

Huge teeth got in the way of more than just running; they made eating and other basic activities difficult. The simple act of ingesting food was awkward because the enormous canines got in the way. Sabertooths had to place their faces sideways against their kills, gnawing food in through the sides of their mouths to get around their massive daggerlike teeth.[14]

Because of the complications extreme size introduces, the weapons of most predators remain small. Teeth, claws, and claspers are sharp and lethal, but not particularly large or spectacular. Such are the canines of lynx: longer than the surrounding teeth and effective for separating the spinal vertebrae of hares, but not so large as to hinder overall agility or head angle in any way, and definitely not large enough to impede the speed and coordination so essential to lynx survival.

Teeth face trade-offs in shape as well as size. One tooth cannot excel at all tasks. Long and slender teeth such as canines are very effective at piercing skin, muscle, or viscera, but these same teeth can snap if they strike bone.[15] Sturdy, bladed teeth, especially if they line up precisely with other sharp and bladed teeth on the opposite jaw, are great for shearing through muscle and sinew. But these blades fracture if they are used to crush or grind bone, and even accidental contact with bone can dull the blades, rendering the teeth useless at their primary function. Alternatively, wide, solid, dome-shaped teeth are great for cracking bone to reach the nutritious marrow, but they are useless at slicing, piercing, or puncturing.

Enhanced performance in one context can detract from performance in another, forcing a compromise. In this case, the inability of a single style of tooth to pierce, slice, and crush has been a fundamental obstacle hindering the evolution of an increasingly specialized predator weapon.

The success of mammals may be attributed in no small measure to the fact that they stumbled on a mechanism to circumvent this trade-off, at least partially. Mammalian carnivores uncoupled the evolution of subsets of their teeth, so that each set evolved to function like a different tool. This way the mammalian jaw could carry three or four tools (for example, canines, molars, and premolars), and each could tackle a distinct task.

This was no easy evolutionary feat, and other types of predators never managed it. For example, theropod dinosaurs, notorious flesh-specialist predators (including *Allosaurus*, *Carnotaurus*, and the in-

Tyrannosaurus rex and other flesh-eating dinosaurs lacked specialized subsets of teeth like molars and premolars.

famous *Tyrannosaurus rex*) had no obvious parallels to molars or premolars, no bladelike edges for shearing, no dome-shaped caps for crushing. Instead, all of their teeth were roughly similar in shape to canines. As a result, even though they did diverge in body size, allowing some partitioning of prey resources, theropod dinosaurs never diversified to fill the breadth of ecological roles seen in carnivorous mammals. In other words, there were no bone-cracking or saber-toothed theropods.[16]

By uncoupling the shapes and functions of particular subsets of teeth, carnivores became incredibly successful and specialized hunters. But this solution was far from perfect, and the fundamental limitations of these trade-offs remain. Canines, premolars, and molars are all still housed side by side in the same jaw, which is rather like opening all the tools in a Swiss Army knife at the same time. This means that their distinct functions can only be utilized by careful chewing, positioning bone over the dome-shaped molars, sinew and meat over the bladelike premolars, and keeping the canines out of the way.

The luxury of careful chewing that we might experience while enjoying a fine steak at a French restaurant is rarely afforded to free-ranging top predators, who face constant and intense competition from rival predators attempting to steal their kills. So the reality is that animals must slash and crush fast, and in this real-world haste, mistakes get made. Blades get worn and teeth crack. A survey of both living and extinct carnivores shows an astonishing frequency of natural tooth breakage, with one out of every four teeth chipped, cracked, or shattered.[17]

• • •

A similar balance between size and performance can be found in the teeth and jaws of predatory fish, especially in the cruising predators of open waters such as tuna and bluefish. Like mammalian carnivores, these animals are often the top predators in their communities, and they can be enormous. Large fish have big jaws and teeth, enabling them to gulp down big prey.[18] Smaller fish with small jaws cannot swallow large prey simply because it's physically impossible to fit them into their mouths. Predatory fish must chase and catch their prey while swimming fast and, like the lynx, these predators often fail. In fact, they miss their mark more than half of the time,[19] placing a premium on body shapes that enhance swimming speed.

In principle, a fish should be able to increase jaw and tooth size without having to get larger overall. Then it could catch disproportionately large prey—possibly even prey bigger than itself—without sustaining the metabolic demands of an especially large body size. Again, the problem is the balance between opposing forces of selection. Jaw dimensions affect individual performance in two contexts: swallowing prey and catching prey in the first place. Larger jaws are definitely better for ingesting bigger and more diverse prey. But they are selected against due to the drag they incur as they're pushed through the water.[20] For many open-water predatory fish, natural selection for faster swimming speed acts in the opposite direction from selection for swallowing larger prey. Because the same fish must accomplish both tasks, the result is functional but unspectacular jaws and teeth, and modestly proportioned weapons.

• • •

As a boy I used to splash through the muddy waters of Bull Run Creek in eastern Tennessee where my mother and stepfather have a small farm, to climb up to the crest of a hill on the far side where our neighbor grew tobacco. Surrounded by sticky leaves, I'd gently push between plants taller than I was and walk down the rows, studying the mounds of exposed earth as they sloped away from the base of each plant. I was scanning for shiny pieces of black obsidian or blue-gray flint freshly cleaned by rain. A notch on one side, or flaked edges that came together at the right angle—some pieces just looked right, and I could tell even from the smallest bits of exposed stone that beneath the surface lay a masterpiece. Of course, most of these "masterpieces" turned out to be nothing; but every once and a while I'd pluck from the dirt a beautiful work of art.

Two thousand years ago some hunter sitting on this Tennessee hill smacked a hammerstone against a fist-sized obsidian "core," shearing off a slender, two-inch flake. Then, knapping gently with the hammerstone, he chipped off small pieces, roughing out the overall shape of the point. Finally, pressing a piece of antler bone against the edges and marching in tiny steps, he popped a precise row of slivers away from each side, working the point until it was symmetrical, with a razor-sharp blade along the length of each edge. The result: a three-fourths-inch-long tip for a lethal and efficient predator weapon.

My neighbor's field was filled with arrowheads. Discarded flakes lay everywhere, suggesting that this flat-topped hill had once been a village where the points were made, rather than a battlefield or hunting ground. Most of the arrowheads were broken, but when I did find an intact point poking out of the mud, I'd close my eyes, sweet-smelling tobacco leaves rustling in the breeze beside me, and squeeze the arrowhead tightly in my hand. The last person to touch it had been the man who made it. For the briefest of moments it felt like I'd touched the past.

Two thousand years may seem like a long time, but even by North American standards the points I pulled from my neighbor's field were young. Stone points and the spears, atlatls, and bows and arrows that delivered them were the primary weapons of humans for tens of thousands of years.[21] Millions have been collected and archived, exposed by plows or eroded from lakeshores and the banks of streams, allowing archaeologists to trace how their shapes and sizes changed through the ages.[22] Remarkably, almost all of these weapons are small. Like the canines of lynx or the jaws of fish, the sizes of stone projectile points reflect a balance between killing power and portability.

As early as fifteen thousand years ago, North American hunters began to throw their spears with a throwing stick, or atlatl, and to tip their spears with points flaked from sharp stone.[23] Spears function well only if their tips are balanced with the rest of the shaft, and if the blade is wide enough to create an incision in an animal's hide sufficiently large for the attached shaft to enter.[24] This puts strict limits on the sizes of tips that will function on particular wooden shafts: bigger shafts require bigger stone points. Large spears are also heavy, hitting their prey with greater force and deeper penetration than smaller ones. Not surprisingly, large spears and points enable the takedown of large prey.[25]

But just as with mammals, the benefits of large weapon size are offset by considerable costs. Larger points require rarer materials such as large obsidian or flint cores with no internal imperfections, and they take substantially longer to make.[26] Large spears are also bulkier, and hauling them around can be tiring. Early hunter-gatherers traveled extensively as they tracked the seasonal appearance of fruits and tubers and the movements of large game—possibly as much as three to six miles a day and more than two hundred miles every year.[27] These nomadic people had to carry their weapons with them as they walked, along with everything else they owned.

The record shows thousands of years of stasis in point size and shape, and when these points did begin to change in size, they actually evolved to smaller, rather than larger, sizes. At least two factors contributed to this gradual reduction: changes in the sizes of prey available, and changes in the technology used to propel the weapon.

The earliest North American points (Clovis points) reach 8 inches in length but were typically closer to 3 inches, and were consistently found alongside mammoth bones.[28] Indeed, the Columbian mammoth

15,000 years of weapon evolution in North America

appears to have been the primary prey species for North American hunters until approximately twelve thousand years ago, when mammoth populations crashed and hunters began to shift their efforts to the giant, and also now-extinct, *Bison antiquus*. These ancient bison were six times smaller than mammoths (only three thousand pounds, on average, instead of eighteen thousand pounds), and the early hunters' appetite for them was associated with a steady reduction in spear and point sizes (Clovis points associated with *B. antiquus* averaged about 2 inches in length, and later Folsom points associated with this same bison species averaged 1.5 inches).[29] When this species, too, disappeared, hunters tracked still smaller prey types, including modern bison (*Bison bison*), as well as bighorn sheep, deer, elk, and antelope, and point sizes shrunk accordingly.[30]

At the same time, several key innovations in weapons technology occurred. The addition of feather fletching to thrown spears around 7600 years ago dramatically improved the speed and accuracy of spears. But these advances worked best with smaller, lighter shafts and correspondingly smaller points.[31] Then, around 2000 to 1300 years ago, the bow and arrow replaced atlatl-thrown spears, favoring still smaller shaft and tip sizes.[32] Hunters could shoot arrows farther and faster with a bow than they could hurl darts with an atlatl, improving hunting success still further and making it possible to kill all sorts of new prey animals.[33] Arrow tips, now much smaller than their Clovis predecessors, could be manufactured rapidly using readily available materials, and the bow and arrow was significantly more portable than a spear.[34]

As with carnivore teeth, a balance between opposing agents of selection kept the stone tools of early humans modest in size. This sort of compromise with weapon size is almost always necessary for predators, and it's why the vast majority of animal weapons are small. Yet, exceptions to this rule definitely occur. In certain special situations, the shackles of balancing selection are shattered. In these populations, weapons begin to get really big.

3. Claspers, Graspers, and Giant Jaws

In the fall of 1992, I reunited with another college buddy for a ten-day excursion to South America. We couldn't afford to go to Machu Picchu, so we settled on Ecuador, a beautiful land I'd explored three years earlier in my search for rhinoceros beetles. The plan was to climb a mountain and then spend several days relaxing by a lake in the rainforest. Volcán Tungurahua was still over eighteen thousand feet—it wouldn't erupt for another six years—and the views from the summit were spectacular. Now, sunburned and sore, we'd bounced for twelve hours in a crowded bus to the frontier town of Coca to meet up with our guides, Clever and Selfo, for the boat trip into the forest. We stashed our gear in the motorized dugout canoe and hopped into the back of their pickup to fetch last-minute supplies.

My friend Chris spoke no Spanish. I could get by, with difficulty, and we'd just discovered Clever and Selfo spoke no English. We weren't particularly worried until they pulled over by the side of the road and peeled a road-killed javelina from the tarmac, hacking off a hind leg and wrapping it in a sheet of plastic, placing it gently in the back of the truck next to the cooler and boxes of food. Chris and I exchanged glances. I leaned into the little window that opened into the cab of the truck and tried my Spanish, asking what the pig was for. "*Cena*" was the response. I was pretty sure that meant "dinner," which was not the response we were looking for. I probed again, but the only word I could discern from their answer was "*cena.*"

Eight hours later we glided into our campsite, having motored more than seventy-five miles down the Napo River and then another dozen up the Pañacocha tributary. There wasn't another person for dozens of miles in any direction. Save for a raised wicker platform for our tents and a table with a cookstove, the place was pristine. By this point the javelina leg was getting ripe, and we again watched with alarm as our guides placed it beside them and began assembling cookware for dinner. Brushing away an assortment of flies, Clever pulled back the matted skin and sliced into the muscle, making a pile of sugar-cube sized chunks, scraping the lot onto a platter, and handing them to us raw. Then he unwound fishing line from a spool in his pocket, gave us hooks, and pointed to the water. The pig was for dinner after all—just not in the way we'd expected!

Piranha fishing is so easy it's almost not even fair. Skewer a chunk of meat onto a hook, toss it into the lake, and wham; pull out your fish. Over and over we cast, and every single toss scored another fish. In no time at all we had two dozen fillets simmering in butter over the fire. To this day I consider piranha the best-tasting fish I've ever eaten, and those nights in the forest we feasted like kings. (We even went swimming in the lake the next morning—exhilarating, now that we knew we were surrounded by flesh-eating fish; the trick, incidentally, was to jump straight into deep water from a boat, rather than wade in from the shore.)[1]

• • •

Not all predator weapons are small, and piranha teeth are a case in point. Piranha jaws are packed with giant, triangular blades that bulge out of their mouths in a fierce under bite, even when their jaws are closed. Piranha don't feed like other fish. In addition to swallowing prey whole, they also rip off chunks of flesh one bite at a time. This simple change in hunting behavior released these fish from the constraint of having to be physically larger than their prey. Now they could swallow small fish as well as parts of bigger fish, including scales, fins, and gouged mouthfuls of flesh.[2]

Popping mouthfuls of meat from big animals required quick lunges from short ranges, rather than prolonged chases through open water. Piranha also scavenge, stripping even human carcasses to skeletons.[3] But scavenging, like lunging, works just fine with awkward jaws. Because piranha no longer needed speed to catch prey, selection favored thicker jaws and larger and larger teeth. Barracuda, incidentally, have similarly impressive teeth for precisely the same reason.[4]

Sit-and-wait, or ambush, predators take weapon evolution to an even greater extreme. Sabertooths were ambush predators who dropped from the branches of trees to plunge their daggers into the necks of unsuspecting prey. Like piranhas, ambush predators no longer chase after prey to hunt them down. In fact, most of them don't run or swim fast at all. Instead, they lurk motionless, often blending spectacularly with their backgrounds like a hunter in a blind, waiting for prey to come to them. When unlucky edibles happen by, these predators lunge from their hiding places, striking out with a snap of their jaws or a flick of their legs to snatch and incapacitate prey before they even recognize what is happening, much less have time to escape.

Deep-sea ambush predators often use lures—dangling globs of light that act like beacons in the immense blackness of the extreme deep. Because prey come right up to these predators, the need for cruising speed is gone. The constraints imposed by drag are minimal, and selection for increased jaw size prevails. Many of these fish, with evocative names such as viperfish, ogrefish, and the humpback anglerfish, have extremely large mouths studded with long, sharp teeth. "Umbrella-mouthed" or "gulper" eel jaws are so big that the fish's body is literally just a giant mouth with a tail. Their gape can be as wide as their body is long. When these eels unfurl their enormous mouths, engulfing water like giant submerged balloons, they can swallow prey that are even larger than they are.[5]

Most praying mantises are ambush predators, which explains their supersized forelegs with long curved spines. These insects get their

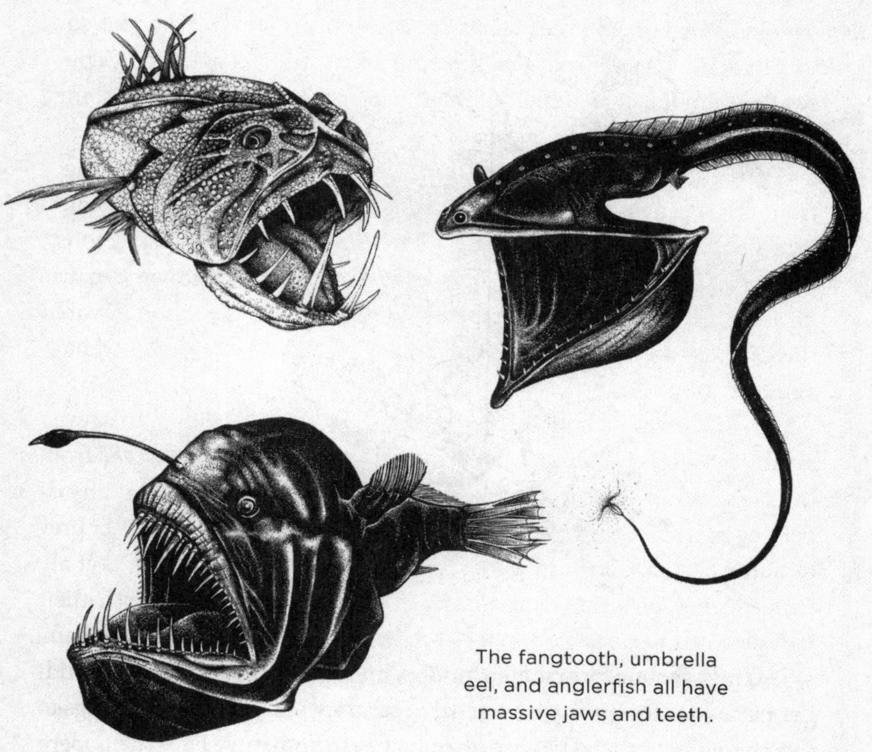

The fangtooth, umbrella eel, and anglerfish all have massive jaws and teeth.

name from a habit of holding their big forelimbs in front of their faces, a posture that resembles a person in prayer. In fact, these long raptorial limbs are spring-loaded with recoil and muscle, and their placement can be compared to the cocking of the hammer of a gun. These toothed limbs snap out from the body, grasping any prey that make the mistake of wandering into the "kill zone."

Early praying mantis species were skinny, generalist hunters who stalked along the ground or through vegetation. Their forelegs were slightly enlarged, facilitating a rapid snatch of spiders and insects they happened across. From these ancestors, mantises became increasingly specialized as sit-and-wait hunters.[6] Once balancing selection for efficient locomotion waned, their forelegs became bigger and bigger, enabling mantises to capture prey at greater distances.[7]

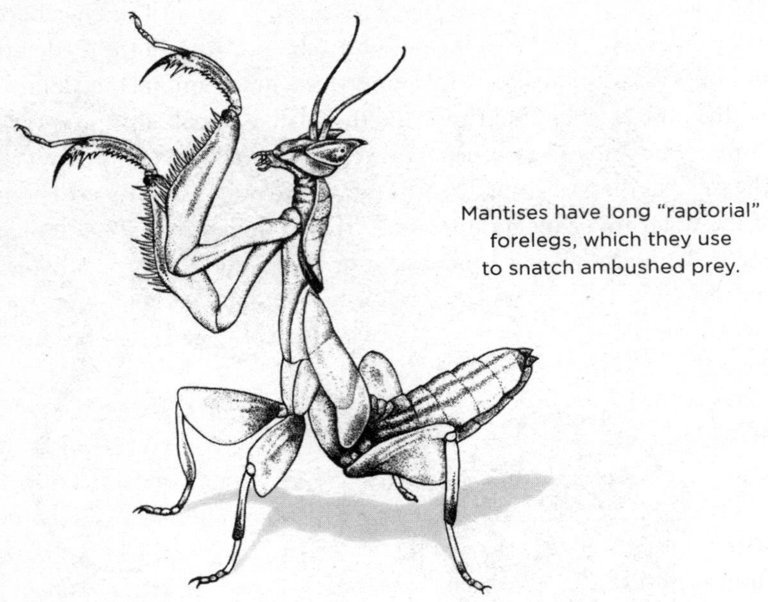

Mantises have long "raptorial" forelegs, which they use to snatch ambushed prey.

Mantis shrimp provide an underwater parallel to these terrestrial graspers. Mantis shrimp are neither mantises nor shrimp, but they get their name from their striking resemblance to both; they have shrimplike bodies and enlarged, raptorial limbs that look like those of a praying mantis. These thumb-sized crustaceans hide inside burrows in rocks or coral on the sea floor and ambush snails, other crustaceans, or bivalve shellfish. They are called "smashers" because they use their distended forelimbs to spear and smash their prey in a highly effective manner: single strikes from these weapons can be deadly. Mantis shrimp hurl their legs with incredible speed during these blows, which is nontrivial considering they are underwater. Sheila Patek and Roy Caldwell, studying the mechanics of the predatory strike of the peacock mantis shrimp, found that they accomplish this feat using a "click" mechanism in which a rigid latch locks the limb into a spring-loaded and cocked position. When this lock is released, the weapon snaps forward using elastic recoil from energy stored in the bending of the skeleton, exactly like releasing the recoil stored in a fully drawn archer's bow. Once released, their legs blaze through the water at strike speeds of up to sixty miles per hour which, when scaled down to the size of these "shrimp," means that the entire strike takes just two one-thousandths of a second.[8] These little predators are contenders for the fastest weapon strikers in the animal kingdom.

The limbs of mantis shrimp (and the related "pistol" shrimp) sweep through the water so fast they create explosive vacuums in their wake. The gaps they leave behind them as they move pull dissolved gasses out of the water into cavitation bubbles. These bubbles release additional energy with a crack when they collapse. This audible pop is both loud and deadly—a noise topping 220 decibels and a flash with temperatures approaching that of the sun (8,500°F).[9] Although the flashes from these cavitation bombs are too small to see, the shock waves they generate can stun nearby fish. The combined impact of an incredibly swift strike, followed by the explosive release of energy from collapsing cavitation bubbles, can crush the exoskeletons or shells of prey instantly.

Other extreme-weapon predators creep along stealthily and surprise prey. This stalking method is similar to sit-and-wait ambush hunting in that it relies less on rapid chases or efficient locomotion and

Alligator gar sweep long jaws to the side to strike prey.

more on swift and lethal strikes from close range. Many stalker fish have extraordinary mouths. Alligator gar, for example, have long, thin jaws packed with razor-sharp teeth that enable them to snatch prey with a sudden, sideways sweep of their mouth. Caimans and crocodiles have similar jaw shapes for precisely the same reason.

• • •

The success of all of these ambush and stalking predators hinges on a quick strike with a weapon that immediately incapacitates its victim. Speed matters, but it's not the sprint speed of the whole animal that determines the success or failure of these hunts. Instead, it's the speed of the appendage that matters—how fast it springs out from the body. Here, bigger weapons are almost always better: longer claspers or jaws can reach out farther from the body; bigger appendages can house stronger, thicker skeletal elements incorporating greater elastic recoil and bigger, faster muscles; and hooks or claws at the ends of these appendages move faster the farther they are from a joint or hinge; longer appendages fly through air or water faster than shorter ones.

The physics behind this last point is called the "law of the lever," and a good way to think about it is to consider a seesaw. If the pivot sits in the middle of the seesaw, then each end of the board moves the same total distance up or down, and each end moves through the air at the same speed (they travel the same distance in the same amount of time). But slide the pivot closer to one end of the board, and two things happen. The distance traveled by the two ends begins to differ, since the long end travels a more dramatic arc than the shorter end. And the speeds of the ends of the board diverge. Both ends of the seesaw complete their respective arcs in the same span of time (presuming that the board doesn't bend). But the long end travels farther than the short one, which means that the long end also travels faster than the short end of the board. The farther an object like a tooth sits from the hinge of a lever, the faster it moves when that lever rotates.

Carnivore jaws illustrate the other half of this principle. Cats and hyenas each sacrifice jaw speed in favor of jaw strength, with squat faces and canine teeth migrating relatively closer to the hinge. Here it helps to think of a nutcracker, or a pair of pliers: the closer an object is placed to the hinge, the more powerful the closing force that can be applied. Wolves, on the other hand, have longer jaws than hyenas or cats. Their canines bite with less force, but what is lost in force they gain in speed. Their canine teeth are situated farther from the hinge of the jaw and, as a result, move faster when the jaw closes.

In ambush predators and stalkers we find this logic carried to an even greater extreme. A specific ecological situation has relaxed the counteracting forces of selection that normally constrain the evolution of large weapons, and selection for snatching predominates.

• • •

Highly social insects, particularly ants and termites, have escaped the weapon size versus locomotor performance trade-off in a different way: through a division of labor. Colonies of these insects can be immense, often with millions of individuals all functioning together as an efficient whole. Part of this colony efficiency arises from a subdivision of tasks among workers with specialized body shapes, not unlike the different subsets of teeth in carnivore jaws.

The ability to uncouple development paved the way for independent evolution of carnivore teeth—canines diverged from premolars, and premolars from molars. In social insects, an analogous uncoupling permitted soldiers and workers to evolve independently, and to diverge considerably in form.[10] Soldiers didn't need to run efficiently, fly, conduct colony maintenance tasks, or even reproduce. They had only to perform as soldiers. Their release from these other tasks meant that the negative consequences of enlarged weapon sizes were minimal.

Within the ant genus *Pheidole* (the "big-headed" ants), for example, individuals fall into one of several very different "castes," including winged reproductive males and females (who disperse from the colony in huge, synchronous, mating swarms), small workers, large workers, and soldiers. Soldiers in these ants have evolved to become fighters with grossly enlarged heads, jaws, and teeth.

In the trap-jaw ant, soldier jaws are long and curved, with sharp teeth. These ants use a lock-and-release mechanism of jaw closure quite similar to the appendage strike of the mantis shrimp. Jaws of this ant can close at speeds of up to 143 miles per hour, snapping from completely open to shut in less than a thousandth of a second.[11] These jaws are so fast that if the ants snap their jaws when their faces are aimed against the ground, they can launch themselves backward twenty body lengths into the air—which turns out to be a very effective escape tactic.

Army ant soldiers have giant jaws and thick, distended heads. En masse, these formidable little warriors can topple scorpions, lizards, and birds. They also can be handy to humans, as at least a few tropical biologists can attest. As a graduate student I spent three weeks in Belize with a biology class, living in the muddy forest understory in tents learning how to conduct field experiments. I wore a machete in a cheap plastic sheath hanging from my belt, and one afternoon as I stripped for a swim the handle of the machete caught on my pants. I must have been talking or otherwise distracted, because I never noticed the blade sliding across my thumb until it was too late—I'd sliced clean through to the bone. We were miles away from civilization with no easy way to get to a hospital. So we sterilized the wound with rum and sutured it with ants. One person held the edges of the cut closed, while another carefully positioned the ants. Angry soldiers reach out with their huge jaws agape, but as soon as you place them against skin, the jaws clamp shut. Pop the rest of the body from the head and you have a surprisingly effective suture. A row of five or six ants worked nicely, once I got used to the idea.

Termites also show division of labor with specialized soldier castes, although these fighters are involved primarily in colony defense, rather than offense. *Incisitermes* soldiers have thick, muscle-filled heads with huge jaws. *Nasutitermes* soldiers have completely different adaptations; they squirt sticky filamentous threads at invading ants. The sticky strings tangle the legs of ants, incapacitating them. Nasute soldiers have no eyes or mouths. Their entire head is one giant, bulbous nozzle with a snout—a walking squirt gun.[12]

Division of labor has always been an integral part of human military forces, too, circumventing the same trade-off between portability and size. Light infantry can move far more swiftly than heavier artillery, for example, and the largest guns are cumbersome and difficult to maneuver. Despite attempts to overcome these limitations, such as placing catapults and cannons on wheeled carts or, much later, housing big guns on rail cars and tracked vehicles, armies have always faced

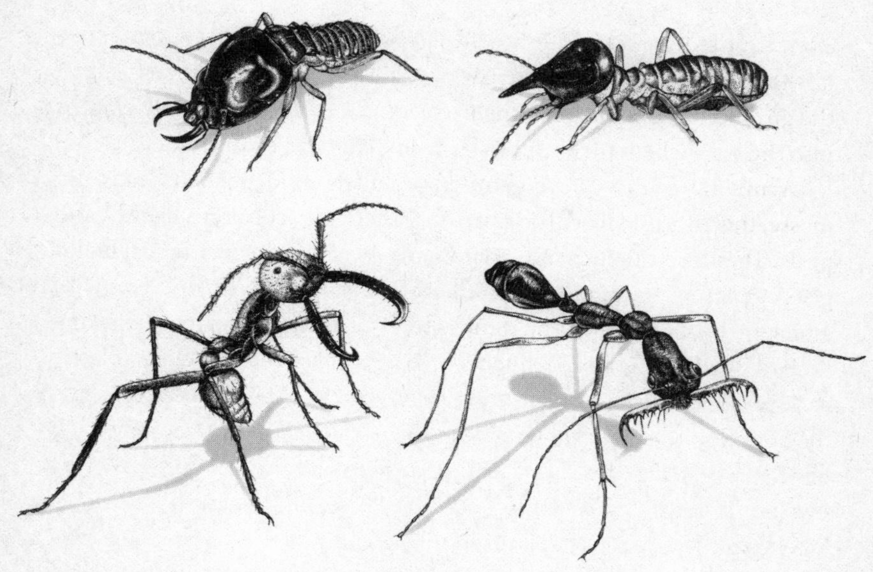

Four soldiers: biting and squirting termites, and army and trap-jaw ants

limitations due to the unwieldiness of their biggest guns.[13] The solution was to separate the functions of infantry and artillery units, permitting one to specialize for speed; the other for firepower.

During the First and Second World Wars, navies faced the same constraint, and they adopted an analogous solution. Battleships got bigger and bigger, packing more and larger guns, but increased firepower came at the expense of speed and maneuverability, so battleships relied on smaller, faster ships such as cruisers and destroyers to screen them and provide reconnaissance.[14]

• • •

In the animal world, extreme weapons are usually confined to specific, rare circumstances such as sit-and-wait hunting tactics, or within-colony division of labor. But another surprisingly potent phenomenon can lead to the evolution of exaggerated weapon sizes: competition. Nature's biggest weapons decide the outcomes of battles over the chance to reproduce, the most precious commodity of all.

PART II

TRIGGERING THE RACE

Specific circumstances must fall into place before weapons launch into an arms race. Appreciating these "ingredients" reveals much about the function and diversity of nature's most extravagant weapons, including why some species have them and most others do not.

4. Competition

Jacanas are bizarre birds, especially when it comes to their weapons. Black slender bodies contrast starkly with yellow bills, long yellow wing spurs, and wrinkled, fleshy wattles—folded globs of feather-free skin that look like pieces of chewed cherry-red bubble gum squished onto their foreheads. The female I'm watching has especially big spurs, one on each elbow, and she creeps about with delicate steps on very long legs. Her slender toes splay to a span of more than five inches per foot, and her gait brings to mind a carnival stilt walker.

This female, red-over-blue-and-white right (named for the colored ring bands on her right leg), is making the rounds of her territory, checking the nests of each of her four mates. Her territory is difficult to get to, and we have to watch at a distance from a canoe. Jacanas defend areas of floating vegetation on wide tropical rivers. Panamanians call them "Jesus" birds because they look like they walk on the water and, in a sense, they do. They tiptoe across their floating mats, dispersing weight with each step over delicate toes, balancing atop the bobbing rosettes of water lettuce and hyacinth. Most predators can't reach jacanas out here, since they'd sink through the thin mat into the river. Crocodiles and caiman, however, swim under the mats and float up into the lettuce, snatching birds from below.

Morning sun cuts through thick mist rising from the Chagres River, the water source for the Panama Canal, and steam lifts from the tropical forest nearby. It is unbearably humid, and biting flies nip at our ankles and the soles of our feet. We crouch in the canoe, trying to get comfortable as we follow our focal birds. I'm using binoculars, sweat already dripping down my arms as they steady my view, elbows planted on the hot aluminum rim of the boat. My father sits behind me peering through a spotting telescope. Mounted on a tripod and perched rather precariously between us, the scope is trained on one of the males, who just this morning hatched four chicks. My dad is a biologist who studies the behavior of birds, and on this morning in 1987 he was beginning a multiyear project in Panama examining the behavior of jacanas. A sophomore in college and eager for adventure, I joined him in Gamboa for a month to help out.

Each morning we drove to the edge of the river where an old Grumman canoe lay chained to a tree. After loading water, lunch, ponchos, clipboards, and binoculars, we'd paddle a mile upriver and cross the huge channel to the far side, where a number of floating mats sat in a wide eddy, as stationary as islands floating on the surface of a river can be. Here he'd ring-marked most of the territorial birds—the ones able to hold on to the precious vegetative real estate—and we spent day after day spying on our avian actors as their lives unfolded on this patchwork of swirling, drifting stages.[1]

This morning, red-over-blue-and-white right is fighting again, as she so often does. An unmarked female had darted in from the adjacent shoreline, hiding in the hyacinth leaves behind one of the males, but our territorial female spots her immediately and closes in. Now, face-to-face, the two birds size each other up. Crouching low, elbow spurs flared out to the sides, each sidesteps the other in a slow circle. Then red-over-blue-and-white right pounces, leaping into the air and striking the intruder feetfirst on her way down, slashing out with her spurs. Everything whirls into a blur as both birds leap at each other over and over, crashing together and jabbing as they flail onto the mat of floating lettuce, pop to their feet, and leap once more. And then suddenly the fight is over; the intruder flies away, and the thick air rings with raucous *ka-ka-ka-kas* as our focal female proclaims victory to birds nearby.

Hundreds of vagrant females forage along the nearby shores of the river. These individuals have failed to secure an island territory of their own, and they challenge the resident females incessantly, pressing and probing, searching for weakness. For vagrants these battles are "do or die," since failure to secure a territory is an evolutionary full stop; a dead end. Unless they can find a way to displace one of the owners, their chances of breeding are nil.

Female jacanas are fighters, towering over the males. They are stronger than males, vastly more aggressive, and they have the larger weapons.

Jacanas are unusual because females have larger weapons than males: a pair of yellow wing spurs.

Sharp yellow spurs jut forward like daggers from each elbow. Bigger females fare better in fierce battles and, as a rule, only dominant, top-condition females manage to hold a territory for long enough to breed.

Males also fight for territories on the floating mats, though their battles are less vicious and are independent of the wars waged by the females. Males fight with rival males, and successful individuals defend patches of floating greenery sufficient to raise a clutch of chicks. These territories pack into the floating islands like tiles of a mosaic, with the female territories superimposed on top. Some females may be able to shove rivals away from only a single male territory, but the biggest and best females own enough island to house three or four males.

The sky cracks with a boom, and warm rain begins to pour down (it does this a lot in Panama). Torrents of water dump over us as we scramble to cover the scope and our notes with ponchos and plastic. The birds hunker down to sit out the deluge. We hunker down, too, and wait, exposed and shivering as lightning crackles around our little metal boat. Ten minutes later the storm has passed and we, and the birds, are back at it. Three inches of water sloshes along the floor of our canoe, so we flip a plastic milk crate on its head and use it as a table to keep our gear clear. My female is engaged in yet another fight—her fourth of the morning—and I check in with my dad. The male he's following shepherds his brand-new chicks as they wobble from plant to plant, all feeding on little insects squirming at the water surface around the lettuce.

The next male over, his territory also nestled within the holdings of red-over-blue-and-white right, still has eggs, and he sits on his hidden nest sheltering his clutch. A third male has nearly grown chicks, and the last male is between broods, ready to begin the process again. Our female, when she's not battling to hold her spot, moves freely among her males, mating with them from time to time. When one of her males is ready, she'll lay a clutch of four eggs into his nest. But then she will abandon the eggs to the male and move on, placing her next clutch a few weeks later into the nest of a different male. Male jacanas spend several months tending to young each time they reproduce, preparing nests, incubating eggs, and shepherding the chicks as they grow to independence. Females show up to provide eggs when needed, but otherwise they leave the tending of young to the males.

• • •

In every way that matters for this book, jacanas are backward. Females are more aggressive than males, they are larger than males, they fight more viciously and frequently than males, and they have larger weapons. Usually it's the other way around. In flies, beetles, mastodons, crabs, and elk, males are armed, not females. Jacanas excepted, in every species with weapons confined to a single sex, males have those weapons. Why should just one sex have weapons? And why is it (almost) always the males?

We have to go back to the beginning for the answer: eggs and sperm. Both sexes offer a copy of their genome to every child, but the packages they come in differ. Eggs are nutrient-rich globs of protein, carbohydrates, and lipids, stuffed into protective membranous cases. Sperm are little more than wriggling, self-propelled packets of DNA. When sperm and egg fuse to launch a new life, it's the resources provided by the mother that first feed it. Billions of cell divisions unfold in a precise cascade of cell-cell interactions as tissues and organs form, and appendages and skeletons grow. All of this takes fuel; proteins to build new cells and tissues, and nutrients and energy to power the literally trillions of chemical reactions. Development is expensive, and eggs provide the energy and nutrients sustaining this process.

Females of all animal species produce larger reproductive cells (called "gametes") than males. Eggs are bigger than sperm, and this difference in material investment is far more substantial than most of us appreciate. Humans are rather ordinary in this respect, but we're a good place to start. The female egg is the largest cell in the human body. It measures almost a fifth of a millimeter across—about the size of a period (.) on this page—and it's just visible to the naked eye. Sperm are the smallest cells in the body, and a hundred thousand could fit into the volume of a single egg.[2]

In many animals differences in egg and sperm size are much more profound. A zebra finch mom fits nicely in the palm of your hand. She's roughly four inches beak to tail, but she lays an egg that is over half an inch across. By weight, a finch egg is 7.5 percent of the weight of her body.[3] That's equivalent to a human female producing an egg weighing eleven pounds. Kiwis have the most gargantuan of all gametes, relative to their body size: brown kiwi moms lay eggs that are a fifth of their body weight.[4] Our human mom would need to produce a thirty-pound egg to compare—the size of an eighteen-inch watermelon.

Asymmetry in gamete size has consequences that ripple through the biology of animal species. For one thing, females can't produce as many gametes as males do. With the same amount of resources, males produce trillions of sperm. And these numbers stack up fast, since *each male* produces similarly copious quantities of sperm. A human female produces roughly four hundred viable eggs over the course of her lifetime. A male, on the other hand, cranks out one hundred million sperm every day, easily four trillion over his lifetime.[5] Scale that up to a population of a thousand people, and there are a quadrillion (that's fifteen zeros) more sperm than there are eggs. Scale it to the current human population and there are a septillion (twenty-four zeros) more sperm than eggs. And humans aren't even an extreme example. The simple fact is that in virtually every animal species there are nowhere near enough eggs to go around. The result is competition.

The size of female gametes is important for another reason. Large, nutrient-rich eggs are expensive, and they take time to produce. Depending on the species, females may take days or even weeks to recover from producing one batch of eggs before they are ready to lay another. Males, on the other hand, tend to need only a few minutes. Thanks to their gametes, females generally take longer to "turn around" between breeding events than do males.

Females also stand to lose more than males if a breeding attempt fails. Although each sex invests nutrients, energy, and time in producing gametes, the amounts they invest are different. Females spend more than males each time they reproduce and, because of this, the cost of abandoning a brood is much steeper for females than it is for males. As a result, whenever additional offspring care is required, it's generally the females who provide it.[6]

Females invest in offspring in all sorts of interesting ways beyond simply producing eggs. Cockroach moms hold fertilized eggs inside their bodies until they are ready to hatch, feeding and protecting them in an insect equivalent of a mammalian pregnancy.[7] Scorpions wear bundles of babies on their backs for weeks after they hatch.[8] Dung beetle moms excavate tunnels into the ground and provision their babies with balls of buried dung; a few species even lock themselves inside a crypt for a year so they can guard the young as they grow.[9]

Preparing nests, pregnancy, guarding eggs, and feeding and protecting young all take time. These forms of maternal care can increase the latency between reproductive events still further, compounding the rift in relative investment between the sexes. Males can invest in offspring, too, as they clearly do in jacanas and humans, but it's surprisingly rare in the animal world. In most animal species, males provide little more than sperm, and this means they recycle an awful lot faster than females.

Turnaround times are tremendously important for explaining animal weapons, because whenever they differ between males and females the result is always competition. If you walk into a population of just about any animal species, and you count how many individuals of each sex are physiologically capable of breeding *right now*, you'll find that all the males are able and willing, but many of the reproductive-age females are not. Some of the females are physiologically unavailable—out of commission, so to speak—in between broods. Female zebra in the midst of pregnancy cannot conceive new foals. Cow elk cannot start new pregnancies while they are nursing existing young. Females that are locked into a current breeding event are "out of the pool" since they are not available for conceiving new young. If all of the males are able to breed, but only a fraction of the females are, then there will not be enough females to go around.[10]

Enter nature's most pervasive and potent form of competition, what Darwin coined "sexual selection."[11] Individuals of one sex compete for access to the other. In principle, sexual selection can work both ways, with either males or females competing. In reality, except for rare cases such as the jacana, it almost always involves males competing for access to females.

Female jacanas still produce the larger gametes (eggs are bigger than sperm), and it takes them a few weeks to recover between clutches. But this is where female investment stops. Male jacanas spend up to three months tending to eggs and chicks and, as a result, their turnaround time is longer than the females' (seventy-eight days, on average, compared with twenty-four).[12] At any point in time, roughly half of the males in a population are tied down with existing eggs or chicks, leaving just a few of them free to start new broods, while most of the females are yolked up and ready to go. They could lay eggs immediately if only they had access to a territory-holding, reproductively ready male. In jacanas, there simply are not enough receptive males to go around, and females battle it out for the chance to breed. Sexual selection, rather than selection from predators or prey, drove the evolution of blazing yellow spurs in jacana females.[13]

• • •

On the hanging scale of parental care (yes, the same one debated by so many busy couples today), the total time invested by jacana dads exceeds the time invested by moms, and the scale tips toward female competition. In some other species, including many birds, the time invested is similar for moms and dads—females lay eggs that are much larger than sperm, but both parents take turns incubating the eggs, and both parents bring food to the nest to feed the chicks. The scale in these cases hangs closer to level, and sexual selection is relatively weak. But these species, too, are relatively rare.

For most animals, parental investment is weighted heavily on the side of females. In one pan sits a tiny sperm, in the other a monstrous egg. Add to that all the time spent preparing a nest; time spent incubating or protecting the egg; time spent nurturing, feeding, and in some cases teaching the young: all of this compounds the initial investment discrepancy arising from the gametes. In these species, the scale tilts sharply to one side, and the arrow points to male competition. The greater the tilt, the stiffer the resulting competition, and the more likely we are to find weapons.

African elephants are a perfect example, and in many ways they represent the opposite end of the investment continuum from jacanas. Bulls invest nothing in the development or post-birth care of their young. All they provide to the females are sperm. Female elephants, on the other hand, have a two-year gestation period and, after birth, they nurse and protect calves for an additional two years. When a female does become fertile, it's only for the exceedingly brief period of five days.[14] Female elephants are only receptive to fertilization for 5 out of every 1,460 days, less than one-half of 1 percent of their lifetime. As a result, there are very few females on the landscape able to breed at any point in time, and way too many males.

Because of this scarcity of fertile females, bulls engage in intense battles with dangerous tusks for chances to mate. Competition among African elephant bulls is severe, vastly exceeding that found in female jacanas. In jacanas, females recycle roughly three times faster than males (24 days compared to 78 days), translating into almost three times as many reproductively ready females as males on floating territories. In African elephants, males recycle more than three thousand times faster than females (less than half a day compared to 1,460 days), and it's not uncommon for many dozens of males to compete for each receptive female. Even at the worst "singles" bars, men don't face odds this bad—possible exceptions being pubs in nineteenth-century mining camps of the American West and, until recently, the bowling alley at the McMurdo research outpost in Antarctica.

Bulls can hear advertisement calls of females—subsonic rumblings carried through soil—at distances of more than six miles, bringing tons of feisty rivals into the fray. Bulls guard females through their brief estrus, but guarding requires consistent victories in the face of stiff challenges, and only the biggest and best-armed males stand a chance. Elephants continue to grow as they age, and elephant researcher Joyce Poole and her colleagues showed that males had to live for thirty years before they even got to play the game—before they were big enough to have any chance of winning fights at all. Most of the time, only males over forty-five actually managed to mate.[15] (Contrast this with females, who generally begin calving by age thirteen). In one long-term study of elephants in Amboseli National Park, Kenya, fifty-three out of eighty-nine males failed to sire any offspring at all, and the overwhelming majority of calves were sired by just three males.[16]

Victorious males wield the longest tusks and tower over their rivals, standing more than twice the height of smaller males. To the victor go the spoils and, in elephants, this means the oldest, largest, best-armed bulls sire the offspring.

Battling bulls

Today, there are just two species of elephant, African and Asian, but not all that long ago a great many species roamed the steppes and plains of Africa, Europe, Asia, and the Americas. More than 170 species have been described, and all but the very earliest had impressive weapons.[17] Columbian mammoth tusks extended sixteen feet and weighed more than two hundred pounds apiece. *Anancus* was a "smaller" cousin to the mammoths, standing only ten feet tall, but the paired tusks of bulls reached thirteen feet apiece. Even African elephants were impressive in their day. In recent decades, poaching and the illegal ivory trade have drastically reduced the size of tusks, and it's rare to see bulls with full weapons. But museums exhibit tusks that are eight feet long and one hundred pounds each, grim testament to the intensity of sexual selection resulting from male competition.

• • •

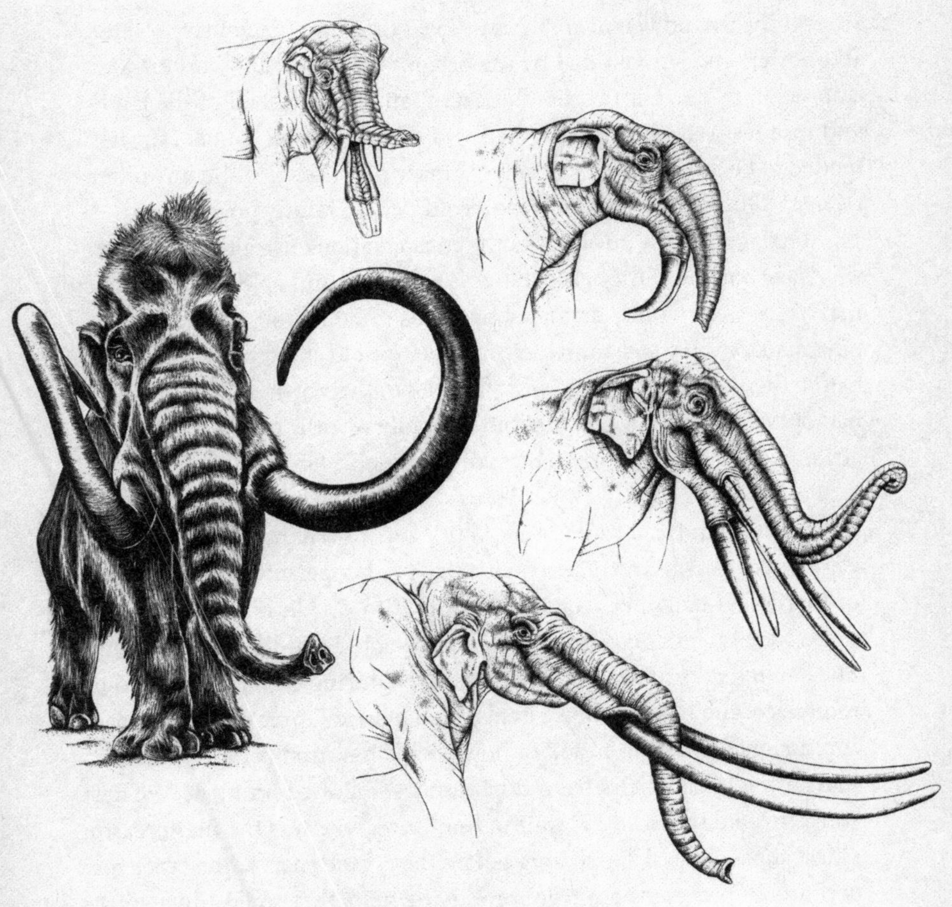

Weapon diversity in relatives of the African elephant

Of course, surpluses of pugnacious young males compete for the attention of females in human societies, too (the reason automobile insurance costs so much more for adolescent men than it does for women), and nowhere was this more apparent than in the European knights of the eleventh and twelfth centuries,[18] when there was a severe limitation of eligible, reproductively ready women, and an excess of battling rival males.

European society during the eleventh and twelfth centuries revolved around locally powerful noble families clinging tenaciously to land and power, and surrounded by masses of tenant peasant workers and laborers.[19] To prevent the dissolution of family wealth, all of the lands and monies passed in their entirety to the eldest son. Noble families tended to be large, with six or seven sons to a household, and any other form of allocation, it was believed, would erode family power.[20]

Marriages, too, were all about the consolidation of wealth and power. Marriage outside of the aristocratic class was unthinkable. Within class, marriages were strictly arranged by heads of households.[21] Eldest sons often had to wait years, until their fathers were old enough to cede power, before they could marry and start families of their own. But at least they had options. They stood to inherit the family wealth, and were therefore attractive suitors for the daughters of other noble families.

For all of the rest of the noble sons—and there were a lot of them—the marital landscape was bleak. With no inheritance to claim, they were not considered attractive matches. A father permitting such a man to marry his daughter would be required to divide his wealth,[22] and there were surprisingly few reproductive-age daughters to start with. The harsh reality of the time was that death during childbirth was commonplace, and heads of households often married three or four times in succession.[23] Existing heads of households had first-priority access to marriageable daughters from other estates, followed, in turn, by eldest sons. Indeed, the need to "wait in line" for a wife was the main reason eldest sons had to delay so long before they could marry (the other reason was to prevent them from producing heirs that could threaten the power of the existing head of the house).[24] By the time heads of houses and eldest sons were accounted for, very few "single ladies" remained.

In fact, the only real option for younger sons of noble families was to marry an heiress—a woman who stood to inherit her family's estate. The violent nature of the day did occasionally result in the total loss of male heirs to a family, in which case a daughter could inherit the wealth.[25] Such a woman, if she chose, could afford to marry a man without an estate of his own, providing him with a chance to found a new dynasty. But noble heiresses were just about as rare as estrous female elephants, and competition for their favors intense.

Noble sons began training for battle by the age of seven, when they signed on as vassals of other knights. They trained incessantly, shadowing their mentors into battle, and learning to run and ride with mail and armor. They could be knighted by the age of fourteen. From that point onward they traveled in bands, roaming the landscape in search of opportunities to demonstrate their valor.[26] There is no question that the primary objective for these men was attracting a wife. Everything they did revolved around besting rival males and, in the process, wooing the favors of noblewomen. Unfortunately, most of the men failed. The few who did succeed in wooing an heiress generally did so after spending thirty to forty years battling rival males and ratcheting their way to the top of the pack.[27]

Actual battle was the best test of a knight's mettle, but it did not lend itself to the attention of women. There simply were not enough battles to go around,[28] so the focus of male aggression turned to tournaments. Tournaments afforded knights with opportunities to demonstrate their strength and courage in front of noble women, and everything about these spectacles reeked of sexual selection.[29] Men fought each other in ritualized battles of strength, often charging from opposing directions on horseback with leveled lances in high-speed clashes that shattered wood and knocked contestants to the ground.

Judges and scribes meticulously recorded outcomes and consolidated results from tourneys across the land, ranking the knights accordingly, and noblewoman studied these standings.[30] Knights emblazoned their armor with colorful plumes and tassels, and bore their family heralds on their shields and breastplates (coats of arms, incidentally, provided information on the genetic quality of a male—his bloodline—not unlike the colorful trains of peacocks).[31] Women inspected the contestants beforehand, watched the battles from front-row seats, and awarded the prizes. And a knight who proved his worth in tournaments could occasionally earn the hand of an heiress.[32]

• • •

Sexual selection differs from most forms of natural selection in ways tailor-made for pushing traits to their extreme. For one thing, sexual selection can be a lot stronger than natural selection. Whenever a small subset of males monopolizes access to large numbers of females, the disparity in reproductive success skyrockets. A few victorious males sire dozens or even hundreds of offspring, while the overwhelming majority of males sire none. When payoffs for success are high enough, weapons can evolve to really big sizes.

Sexual selection also tends to be more consistent than other forms of selection, and this, too, can push traits along the path to extreme. When dark mice moved into coastal beaches, mismatched mice got hammered, and natural selection favored a shift from dark to light. Had we sampled beach populations shortly after the dark mice moved in, we would have measured strong selection acting in one direction, pushing the population toward lighter and lighter fur.

But this pulse of evolution was brief. As soon as beach populations had lighter fur, the directional change ceased because mice now matched their surroundings—any lighter and they would be too white; any darker, and they'd stand out as they did before. Natural selection on fur color stabilized. It reached a balance, and populations now hovered around this new value for the trait.

Such is the nature of most natural selection. Populations adapt to surrounding conditions until they reach a local optimum. Environments may change and, when they do, selection kicks in once again favoring new trait sizes or colors that work better in the new surroundings. But these shifts also are expected to stabilize. Ocean sticklebacks have sported three long spines and fifty-two armor plates for hundreds of thousands of years. When some of these fish found themselves dumped into freshwater habitats, natural selection in their new environment led to a rapid shift from three spines to one, and from many armor plates to few (fourteen plates).[33] But once these populations reached their new optimum, the change in armor stopped.

Even when natural selection is directional, its effects often cancel so that the net effect is still stasis. When physical environments change, they tend to fluctuate back and forth. Winter gives way to summer, but reverts back again each year to winter. Wet years are offset later by dry years, and vice versa. Even glacial ice sheets advance and retreat in cycles, and ocean levels rise and fall. When animal populations evolve in response to these changes, they do so in an oscillatory fashion. Mice become lighter, then darker, and then lighter again. Many populations adapt continuously to changing patterns of natural selection, but the directions of these changes cancel so that the long-term trend is stasis.[34]

This is not how sexual selection works. In battles for access to reproduction, males compete with rival males. The environment that matters for performance is a social one—other males with whom a male does battle—rather than temperature or sea level or other physical features of the landscape. And this social environment evolves in tandem with the weapons. As antlers or horns become bigger, so, too, does the standard against which a male must contend. Like a sliding scale, each increase in weapon size resets the baseline of the population, selecting, in turn, for yet another increase.

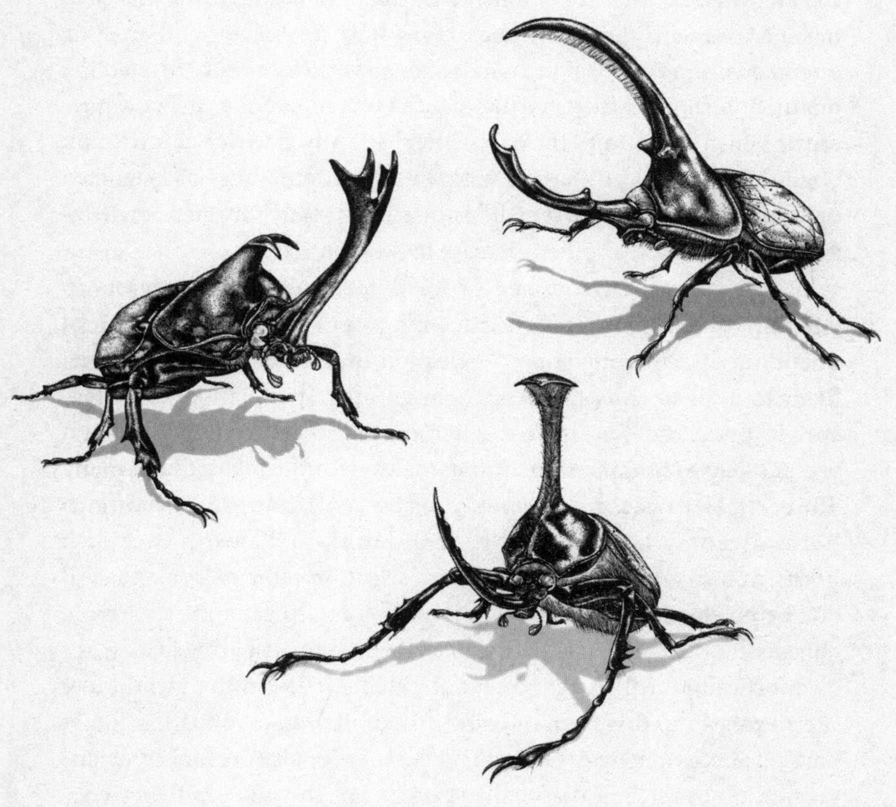

Rhinoceros beetle horns

Imagine a population of rhinoceros beetles where the average length of male horns is one-half inch. In this social environment, a few males stand out. A mutation increasing horn growth has given them three-fourths-inch horns. These males win battles the most frequently; they mate with the greatest proportion of female beetles; and they populate later generations with disproportionate numbers of their kind (including sons wielding three-fourths-inch horns). Over the next few dozen generations, the population shifts. It evolves in response to sexual selection so that now the average size of horns is three-fourths inches.

The benefits of a three-fourths-inch horn aren't so great anymore, because now everybody has horns this big. The evolutionary increase in horn length reset the standard for male competition. Into this new social milieu another mutation arises, this time leading to a one-inch horn. Males with these new alleles now have longer horns than their opponents, and they begin to win. So the new alleles sweep through the population as the largest horned males outcompete the earlier, three-fourths-inch rivals, until the population has evolved to this new weapon size. The scale has shifted once again. The population has ratcheted up to the new norm, and it stands poised and ready for the next mutation leading to still another increase in weapon size.

Because social environments evolve in tandem with increases in weapon size, sexual selection can push populations along a path of unending directional change.[35] Selection in this social context is not likely to oscillate. Measure selection in a beetle population now, measure it in a decade, measure it one thousand years in the future, and you're likely to find the same thing: males with the biggest horns win. The particular weapon size performing best will change (one-half-inch horns give way to three-fourths-inch horns, and then to one-inch horns, and so on), but the direction of selection remains constant. All else being equal, this form of selection is going to generate a lot more change than selection oscillating back and forth or holding fast.

Most of the truly gargantuan armaments in the animal world owe their excesses to this form of competition. Reproduction is the "other half" of success, in an evolutionary sense (we've already looked at survival). But it's the half that really matters. When you strip the essence of life to its core, the only reason to survive is to have a shot at breeding,

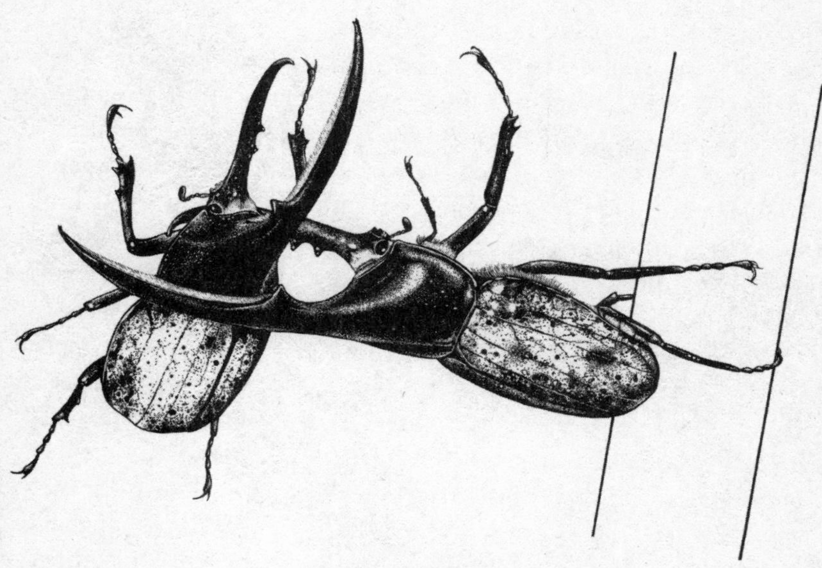

Battling beetles

and, at the end of the day, it's how many offspring you produce that determines success or failure on the evolutionary stage. Individuals reproducing the most win, plain and simple. They contribute more copies of their genome to subsequent generations of the population than other individuals do. Their alleles persist, while others gradually disappear.

Sexual selection arises whenever individuals of one sex differ in their reproductive success. Some males sire three offspring while others sire four; the result is selection for traits that help an individual sire four. But selection in this case will be relatively weak, since the difference between winners and losers is small. When differences in reproductive success get larger, the intensity of selection increases. In extreme populations sexual selection is so severe that it eclipses all other forms of selection acting on these animals.[36] Nothing else matters—not feeding, physiology, or immunity to parasites or disease—and males sacrifice everything for a chance to win the reproduction race.

5. Economic Defensibility

Structures of sexual selection are legendary for their extravagance. But not all of these traits are weapons. Sometimes there is nothing for males to guard; they cannot gain access to females through fighting, so weapons are useless. In these species males compete indirectly, scrambling for the attention of females by wooing them with dances, songs, or bright, gaudy displays. Male túngara frogs broadcast their location with eerie, incessant calls, ringing whines and chucks that are both energetically demanding and dangerous. Males sing for hours upon hours, night after night, even though their calls attract predatory bats.[1] The odds that they will be eaten are high. But the price of failing to attract a female is even higher, so males take their chances and risk death.

Male birds of paradise grow long and colorful tail feathers, which they spread in stunning displays to females.[2] They, too, are risking death, since their elaborate trains make flying clumsy and, as they strut their stuff, they stand out from their backgrounds in full view of predators. Yet they dance recklessly, bobbing brightly colored plumes and shrieking loudly in a biological equivalent to flashing neon signs. Despite the immense risk, males pour everything they have into desperate attempts to outshine the next male. Reproduction is the currency that matters here, and getting picked by a female is the only shot males have at keeping their genetic heritage from vanishing into the abyss of history.

This form of sexual selection is called "female choice" because females actively pick particular males based on the attractiveness of their displays,[3] and it can be just as intense and unending as selection from male competition. Female choice also creates an evolving social backdrop because what is bigger or brighter or flashier depends on the displays of everyone else. Here, too, the baseline ratchets up as each new increase in trait size changes the social context and shifts the standard for what females find attractive. The difference is these extreme traits are ornaments, rather than weapons.

In a sense, males compete with rival males regardless of whether sexual selection proceeds through female choice or male competition and, in many respects, the process—the intensity, consistency, and social nature of selection—is the same regardless. Why, then, do some species embark on a trajectory of overt competition leading to the evolution of weapons, while others end up dancing or singing with displays? Here is where the ingredients for arms races come into play, for these are the pieces that must fall into place if sexual selection is to trigger evolution of extreme weapons. The first ingredient is competition, the essence of all sexual selection. The second is economic defensibility.

• • •

Beetle horns are impressive things. They form during metamorphosis when parts of the body wall grow into rigid protrusions that stick out from the body proper. Horns can be curved, straight, broad, or branched, depending on the species. In many ways they resemble the antlers of moose or elk. Like antlers, beetle horns are typically a male trait. And, like antlers, horns in the largest males can reach enormous proportions. Sometimes a horn will comprise as much as 30 percent of the body weight of a male. If you were to scale these animals up to human dimensions, this would be the equivalent of producing a pair of arms or another leg and wearing it on your head.

Horns have cropped up in all sorts of beetles, from knobby fungus beetles and weevils, to flower beetles, harlequin beetles, rhinoceros beetles, and dung beetles. Dung beetles, in particular, are noteworthy because only about half of the species produce horns, raising the question of why just some species have weapons. There are entire dung beetle genera with hundreds of species that lack horns completely, and other genera that all have horns. I particularly like the genus *Onthophagus* because it contains both. *Onthophagus* turns out to be one of the most species-rich genera of any living thing anywhere, with almost two thousand species cataloged and another one thousand waiting to be described. What is most remarkable is the diversity of horns that they produce and, in this genus, even closely related species can differ so that one species has horns and the other does not.[4]

The heart of dung beetle diversity resides in Africa, where eight hundred of the described *Onthophagus* species are native.[5] In the Great Rift Valley, for example, gazelles, waterbuck, buffalo, giraffes, and elephants, as well as spectacular migrating herds of wildebeest and zebra, all contribute to the copious production of dung. Not surprisingly, dung beetle numbers in these areas can be very, very high.

After weeks spent securing the necessary permits, I finally had a chance to look for dung beetles in Tanzania in 2002, while coteaching a field course through the University of Montana. With armed guards standing on the roof of our vehicle spotting for lions and angry Cape buffalo, I'd rushed from the truck with gloves and a spade to poke through manure as fast as I could. In this way, I was able to sample beetles from buffalo, gazelle, and giraffe dung. But the real find that day was a pile of fresh elephant droppings in the middle of the road. We must have just missed the elephant—the pile lay steaming between the rutted tire tracks. I promptly did what any field biologist would do. I scooped it into a plastic bin and brought it back to camp. That evening, I placed the dung onto moist soil away from our tents and sat back with a ring of students and headlamps to see what would happen next.

Never in twenty years of traveling the world in search of beetles have I witnessed an abundance of insects like I did that night. When they first arrived I tried to catch them, count them, and place each in a separate vial, but soon they came too fast. Even with five helpers beside me all attempting to catch and record these animals we couldn't keep up. Beetles began to spin into our headlamps and plonk onto clipboards and into our laps, and everywhere they fell like rain from the sky. They began dropping by dozens at once. Beetles were tumbling into our hair and down the backs of our necks, and the mass of beetles that landed on my notepad was so thick I had to sweep them aside to write. It was as if somebody was pouring beetles from a bucket over our heads and onto the dung. Our best (and admittedly rough) estimate was that more than one hundred thousand beetles converged on that single pile of dung that night.

Tourists may flock to the Serengeti to see lions or elephants or gazelles with twisted horns, but it is the tiny dung beetles who display the most dazzling weapons here. Just in that one sample, we observed all sorts of impressive weapons: long, forked horns curling up from the head; a single horn rising from between the eyes and arching over the back a full body length beyond the end of the beetle; a long cylindrical horn protruding from the thorax and bending up at the end like a coat hook; and species with one, two, five, and even seven different horns simultaneously.

There were also many species with no weapons at all. Same season, same habitat, even the same pile of dung, yet some species invested in large weapons while many others did not. Why is it that only a few species produce these elaborate structures, given that so many other species do not? It turns out that in essentially every animal species with extreme male weapons the explanation is the same, and it comes from the logic of economics.

• • •

Nature is the ultimate economizer, relentlessly culling individuals who allocate resources poorly. Over time, we expect that populations will evolve to become increasingly efficient in their use of resources, investing in the growth of large structures such as weapons only when the benefits of having them outweigh any associated costs—that is, when they are cost-effective. By any measure, weapons are costly. They are expensive to produce, and they are expensive to use. Males incur risks when they fight, and fighting requires time and energy that could be expended on feeding and other life tasks. However, males with the largest weapons may derive significant reproductive benefits if these weapons help them secure access to females and keep rival males away. When the reproductive rewards are high enough, even extravagant weapons can be cost-effective.

But under what circumstances will the benefits of investing in weaponry outweigh the associated costs? And when will the net benefits (benefits minus costs) of a weapon be the most profound? For many animals the answer depends on the kinds of resources that they exploit, and how easy these resources are to defend.

Imagine a food resource spread out uniformly across a landscape—grass, if you're a grazer, extending as far as the eye can see. As a male, where would you stand guard? Even if it were absolutely necessary for females to visit places with grass to feed, and even if they were willing to mate with you if you happened to be there when they fed, where would that location be? For food resources broadly distributed in space, there is no obvious location that is better than any other. It would be impossible for a male to anticipate where females were likely to visit since they could find their food anywhere. A male could still invest in the production of weapons, and he could use his weapons to deflect rivals from a patch of turf. But why should he bother to guard one particular area if all of the surrounding areas are just as good? And why should he pay the price of producing a weapon and fighting, if other males without weapons or territories do just as well as he does? In the parlance of economists, such behavior would not be cost-effective.

If, instead, those same food resources were sparse, and especially if they were clumped into rare but concentrated patches, then the male would face a very different set of payoffs for guarding a territory. He would still pay a price for producing the weapon, and for expending time and energy fighting to keep rival males out of his territory. But now these territories would matter, and the benefits he could glean from guarding them might be significant. Females would be much more likely to visit him, since the resources they needed were few and far between, and one of the only places where they could access them was inside his territory. Indeed, lots of females might visit his territory and, if the resources were localized enough to be readily defended from rival males, he might be able to mate with a disproportionate number of these females—more females than other males who were not able to guard a territory, and possibly even more than males who *were* guarding a territory, if his territory were bigger or better than theirs or contained more of the limiting resources.

What this exercise in the logic of economics reveals is a central tenet of the field of animal behavior: animals benefit most from fighting to guard territories when those territories contain valuable and limiting resources. The more limiting they are, and the more economically defensible they are, the higher will be the payoffs for successful guarding behavior.[6] But here is where things get interesting, because whether or not a particular resource is valuable enough, or sufficiently localized to make its defense cost-effective, depends entirely on the perspective of the animal in question. What is defensible and valuable to one species may not be to another, and understanding what is valuable to each species is essential in unlocking the mysteries of their weapons.

• • •

Harlequin beetles win my vote for the world's most ungainly animal. They derive their name from angular streaks of orange, brown, and black on their wing covers and bodies, but their most distinguishing features are their weapons: a pair of massive, chopstick-like forelimbs which, in the largest males, can reach to a span of almost sixteen inches. These males are so awkward that when they fly, they have to pull their forelegs back over their heads to keep them out of the way, and their bodies hang vertically as they whirr through the air in slow motion. If crabs could fly, this is what they would look like.

Harlequin beetles are active during the rainy season in lowland tropical forests of Central and South America. David and Jeanne Zeh studied them in French Guiana and in Panama, where their life cycle is intertwined with that of the fig tree locally known as Higuerón. Figs

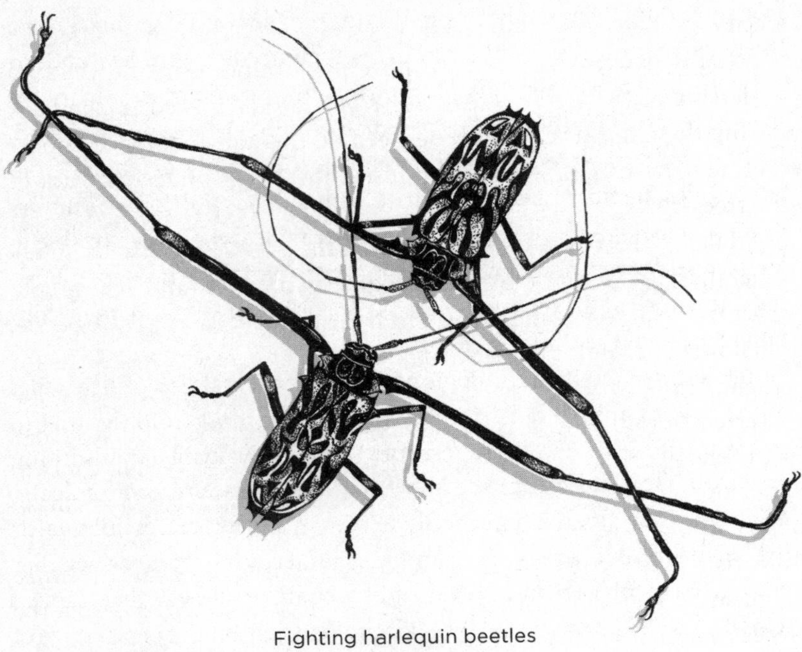

Fighting harlequin beetles

are among the true giants of neotropical forests. Reaching heights of 130 feet or more, these pale trees are easily recognized by their sticky, milky-white latex sap, and by the dozen or more swirling root buttresses spiraling out from their base.

Female beetles drill their eggs into freshly fallen trees, and the larvae feed on the decaying wood as they grow. The problem is that larval development can take a long time—up to a year or more—and this means that only the largest, thickest tree trunks will suffice. Only a collapsed giant will persist on the forest floor long enough for a female's larvae to grow to full size. Fig trees like this don't fall every day.

When such a giant collapses, the beetles flock to it fast—drawn from miles away by the pungent scent of sap released during the cataclysmic ripping of wood and bark as the trunk snaps from its roots and crashes to the ground. A treefall like this tears a gash in the canopy of the forest, letting direct sunlight stream in. As the beetles arrive, they avoid this glare and crowd instead onto the shaded underside of the trunk. A fallen tree still sits on its branches, so that much of its length is actually perched several feet above the ground. Here, on the cool underside of the trunk, sap oozes from slashes nicked in the bark during the fall. Harlequin beetles feed on this sap and, more important, females use these gashes to insert their eggs under the bark of the tree.

Male beetles battle for ownership of these prized spots. There are generally only one or two good territories per tree, and the nearest other treefall can be miles away. From the perspective of a male beetle, sap flows on a fallen fig are rare and defensible—perfect real estate for a fight. Males battle viciously with one another for possession of these inverted territories and the consequent opportunities to mate. They butt heads and grapple with outstretched arms, trying to hook and flip their opponents off the trunk. They rear up on hind legs and smash into each other, twisting and prying with their ridiculous forelimbs in surprisingly athletic scrambles, and slice at flailing limbs and antennae with sharp mandibles. After as much as a half hour of thrashing, the loser eventually falls to the ground, though often not before losing a chunk of leg or antenna in the melee.

To understand why males have such long forelegs, the Zehs needed to observe who won these fights and determine whether winning a fight brought with it opportunities to mate. Their problem was that most of the fighting and mating happens at night. In order to watch the beetles, the Zehs had to hike into the rainforest in the dark.

Nightfall happens suddenly in the rainforest, and getting caught after sundown without a light can be a harrowing experience. The understory is dim even at midday; at night it becomes profoundly dark—with the eerie, disorienting blackness of a cave or a closet where you can't make out even your own fingers. I know people who've had to spend the night on the trail, stuck where darkness caught them without a flashlight. Worse, they had to stand up all night. You can't sit or lie down because the leaf litter comes alive with stinging bullet ants, scorpions, tarantulas, and bushmaster and fer-de-lance snakes. Bullet ants, so named because their sting is as painful as a gunshot, climb tree trunks to forage in the canopy, so you can't lean on a tree, either. Balance gets tricky because you have no visual sense of up or down—no rising bubbles such as divers use to orient themselves—and it becomes hard to tell if you are tipping or not. After a few hours you begin to hallucinate and see light—or so you think—but it is actually the barely discernible glow of bioluminescent fungi and the occasional glowing click beetle waddling through the litter.

When equipped with proper lights, however, the forest at night is exhilarating and wild, full of noise and wonder. To watch the harlequin beetles, David and Jeanne Zeh hiked in to their treefalls at dusk and then put red acetate filters over their headlamps. Because beetle eyes cannot detect red light, the modified lamps did not disturb them, and the red glow was enough for the Zehs to see what happened on the upside-down stage. By painting numbers on the backs of the beetles, they could keep track of each individual and watch many different encounters. They staged fights between males as well. The Zehs observed that males with the longest forelegs almost invariably win, and males who win are the ones who mate with the females[7]—exactly what we expect when sexual selection favors the evolution of increased fighting ability and ever-larger weapons. In short, the extreme paucity of suitable egg-laying locations, combined with the restricted size and ease of defense of these resources, creates in this species an ecological situation where the benefits of fighting and possessing large weapons are enormous.

• • •

The best part of the harlequin beetle story is actually not the beetles. It involves an even stranger animal that hitches onto the beetles for a ride. The Zehs noticed that nestled on the abdomens and tucked beneath the wing covers of their beetles clung tiny little arthropods called "false scorpions," or "pseudoscorpions." This particular species of pseudoscorpion feeds on fallen fig trees, and their nymphs develop inside the rotting wood. Like their beetle counterparts, male pseudoscorpions have weapons—a large pair of clasping appendages called "pedipalps"—and they use these claspers in fights with rival males over opportunities to mate with females. Their weapons, too, are impressive in proportion: much longer in males than in females, and extreme in the largest of males. But in absolute terms these animals are minuscule, and as a result, the context of their battles is different.

Pseudoscorpions battle on the backs of harlequin beetles.

To a giant beetle with a reach of sixteen inches, a wound on the belly of a fallen fig tree is both small and amenable to vigorous guarding. To a male pseudoscorpion, whose reach in the best of circumstances may span just a quarter of an inch, the same wound on that tree is enormous. An aggressive, territorial male may keep rivals away from one side but, while he does this, ten other males can approach females from other sides. Like trying to defend a lake from a spot on the shoreline, the effort is futile; investing in weapons for such fights would be a waste.

However, female pseudoscorpions depend on something else as well, and this resource makes a much more effective choke point, or bottleneck. Fallen fig trees can be miles apart, and these little arthropods lack wings. To travel from one tree to another, they clamber onto the backs of the harlequin beetles and ride from tree to tree. Many pseudoscorpions hitch rides on other insects by grabbing a leg with their pincers and dangling. The harlequin beetle–riding species has a relatively comfortable ride since they have the entire back of a beetle to cling to. They nip the rear of the beetle with their claws and, when the beetle squirms in response, hop aboard. They even spin silk nets to keep from falling while the beetle flies.

It turns out that the back of a beetle makes a perfect mobile mating territory if you happen to be a pseudoscorpion, and males fight with one another to guard this resource. Males use their clawlike weapons in these fights and, as in the case with beetles, males with the biggest weapons win. These fights were understandably harder for the Zehs to

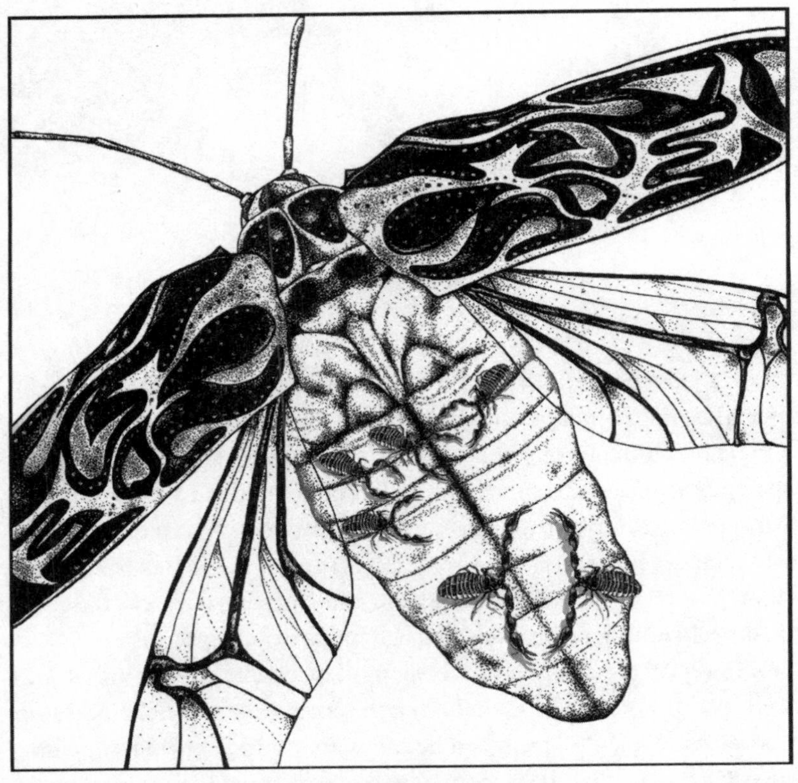

Harem of beetle-riding pseudoscorpions

observe, so they used a method of DNA fingerprinting to identify which male pseudoscorpions fathered the nymphs produced by riding females. They found that males with the largest weapons were the ones most likely to be on the backs of beetles. They also showed that not all beetle backs were equal. More females could ride on the largest beetles, and pseudoscorpion males with the largest weapons guarded these bigger beetles. A successful male may be able to mate with two dozen or more females before the beetle lands. Even better, once they land, the mated females hop off, and a new round of females hop on.[8]

• • •

Harlequin beetles and pseudoscorpions each have something of critical importance to females that is rare and localized, and therefore economically defensible, though the specific resource in each case is different. Males of both species who are successful at guarding the resource are able to mate with many different females. That is, they translate their success in fighting into success in reproduction. And, in both cases, this blend of ecological circumstances has led to a history of strong sexual selection for large male weapons.

Armed with this insight, we can return to the problem of horned and hornless dung beetles. Why do some species produce horns while others feeding on the same resources in the same habitats do not? The answer, I have come to suspect, has nothing to do with the dung and everything to do with what these animals do after they arrive at the dung. The world of the dung beetle is a world filled with intense competition (think of the numbers of beetles that flew into just one pile of elephant dung). Dung is actually a very valuable resource, if you happen to be a beetle or a fly. Rich in nitrogen and other nutrients, it's ambrosia for larvae, and adult insects compete vigorously to get it for their offspring. Beetles must find the dung fast, and then they must contend with the hordes of other insects that also have found it and are attempting to steal it for themselves.

Most dung beetles are either "rollers" or "tunnelers." Rollers are what everybody thinks of when they think of dung beetles, *if* they think about dung beetles. These conspicuous scarabs push their spherical globes of sculpted manure over the ground and tussle with one another along the way. When dung falls onto hard-packed, uncluttered soil or baked clay, they can roll their balls surprisingly fast, and many will travel tens of yards before they stop.

Ball rolling is a superb strategy for pushing food away from rivals. Carve a hunk from the pile, sculpt it into a smooth-sided ball, and roll it away from everybody else. These tasks can be accomplished in minutes, and, by pushing their dung ball several yards away, rollers can escape most of the competition. Ball rolling is most often performed by males, but females will join males as they leave the main dung pad and either cling to the ball and somersault along for the ride or simply follow the male until the pair reaches an adequate patch of soft or moist soil. Here they stop and cooperate to bury the ball, laying eggs beside or on top of the buried dung, depending on the species.[9]

Females are not the only ones to approach males as they roll their balls away. Rival males constantly challenge one another over ball ownership, and vigorous battles are commonplace. But these fights occur out in the open on the exposed surface of the soil. Furthermore, the objects of these fights—balls of dung—are themselves mobile and malleable. Balls are pushed, pulled, even torn in half during fights as the males tussle around and around, clinging to and scrambling over the rolling balls. (These battles are great fun to watch, incidentally. At the Barro Colorado Field Research Station we'd paint numbers on the backs of rival males and place them on a dung pile centered in the bull's-eye of a horizontal dartboard, betting on the winner and cheering them on as they fought to roll their marble-sized balls out of the ring.) Despite their pugnacity, not one of the thousands of ball-rolling dung beetle species has horns.

A second strategy adopted by many dung beetles is tunneling. Females of these species fly into dung and immediately begin to excavate burrows into the soil below. Once they've dug sufficiently deeply—a foot to a yard, depending on the species—they begin pulling pieces of dung down into the tunnels to stash them away from the other dung-feeding insects above. Females may make fifty or more trips to bury sufficient dung to provision just a single egg, and they'll repeat the process for a string of successive eggs. While females are working on this arduous task, male beetles fight among themselves for tunnel ownership. A victorious male

Ball-rolling dung beetles fight in scrambles, rather than duels.

will guard the entrance to a tunnel—not so much to keep other species away from the food as to keep rival males of the same species away from the female. While in residence, a male will mate repeatedly with the female, but he will often get kicked out of the tunnel by an intruding, larger male. Males of tunneling species often have horns.[10]

Tunnels are localized and confined—exactly the sort of fixed, economically defensible substrate in which we would expect to find a performance advantage to large weapons. Males stab their horns into the sides of tunnel walls to block entry of intruder males, or use them to twist or pry opponents loose before pushing them up and out of the hole. In these fights, males can lock themselves into position by bracing against the tunnel walls with barbs, teeth, and thick spines on their legs. Anchored, they can bring the leverage of their weapons to full effect, and males with longer horns win these contests.[11] Because mating almost always occurs inside tunnels, winning fights is a critical prerequisite for reproduction.

Ball rollers have no such leverage. Their fights occur out in the open and over a mobile resource. Although males fight briskly with tumbling and pushing and scrambling, they cannot brace themselves into position like the tunnelers can. As a result there appears to be no leverage or other performance advantage to having large weapons and, without the performance advantage, weapons are not cost-effective. A simple change in the way these beetles hid their food resource had profound implications for the evolution of their weapons.

We've identified two of the three critical ingredients for arms races: intense competition, generally arising among males as they battle for access to females, and ecological situations that cause resources to be localized and economically defensible. There's one final ingredient, and it involves the details of the fights themselves—the way that males face each other in battle. Males must face each other one on one, rather than all together in a scramble. For bigger weapons to perform better than smaller ones, the battles must be matched and "symmetrical," with comparably armed contestants challenging each other face-to-face. Oddly, this last ingredient has been almost entirely overlooked by biologists. To appreciate its significance, we must turn instead to attrition models of military forces, and to the century-old insights of an eclectic automotive and aeronautical engineer.

6. Duels

By the end of the nineteenth century, Frederick William Lanchester had emerged as a brilliant designer and builder of cars. He invented one of the first self-starting devices for gasoline-powered engines, and he was among the first to incorporate carburetors. He'd built his first full car by 1895, and in 1899, he and his brother started the Lanchester Engine Company, one of the first factories in England to build and sell cars to the public. When his company went bankrupt a few years later due to the incompetence of its board, Lanchester shifted his focus to aircraft. He modeled lift and drag for a variety of wing designs, and his "circulation theory of lift" became the basis for modern airfoil theory.[1]

During the First World War, Lanchester started using mathematics to try to predict the outcome of battles. He was obsessed with aircraft, and convinced they could play a critical role on battlefields. In the process of writing his book *Aircraft in Warfare: The Dawn of the Fourth Arm* (1916), he derived a simple set of equations portraying the losses of forces under general circumstances of combat.[2] These little equations, called "Lanchester's laws," sparked an explosion of research into the dynamics of military engagements. Books were written about the influence of these equations,[3] and international conferences were convened to debate their applicability to battle.[4] Although modern models of force depletion in warfare are vastly more complicated than Lanchester's original equations, his models formed the backbone for literally hundreds of subsequent theoretical approaches, and are credited with spawning what would eventually become the multibillion-dollar, intellectually thriving field of operations research.[5]

The logic behind Lanchester's equations was to find an explicit way to calculate how rapidly the forces of one army would be depleted by fire from another, and vice versa. Each army was assigned a force strength, the number of available troops, and a force effectiveness, which translated shots fired from one side into losses incurred by the other. Effectiveness could mean different things, depending on the nature of the engagement and the types of weapons, but it was essentially a measure of the fighting ability of each member of an army.

When armies faced each other in battle, the losses by one side could be calculated as the number of soldiers from the opposite side times the effectiveness of each of those soldiers—for example, the number of bullets fired multiplied by the probability that each bullet hit a soldier from the opposing army. More soldiers meant more rifles and therefore more shots fired (greater force strength); better training, aim, and more powerful guns meant a higher probability of incapacitating a target with each bullet (greater force effectiveness). The objective of the models was to simulate salvos of gunfire by simultaneously calculating losses for both armies with matching equations, adjusting force strength based on these losses, and then repeating the process with additional salvos. Iterated over a succession of salvos, Lanchester's models elegantly revealed the rate of depletion (attrition) of troops and predicted both the duration of the battle and the eventual winner. Permutations to these simulations could then be run with all sorts of altered conditions, revealing the factors leading to victory in each case.

As Lanchester worked through his models, he discovered that out of all the countless advances in types and styles and sizes of weapons that punctuated the history of human warfare, one change altered the rules of engagement more dramatically than any other: the deployment of long-range weapons such as guns and artillery. The way that troops killed each other in the past, when soldiers faced up man-to-man and clashed in combat hand-to-hand, was fundamentally different from the way they killed each other post-firearms. To incorporate these differences into his equations, Lanchester derived two types of models.

The first was designed with ancient battles in mind. Lanchester recognized that when soldiers fought with close-range weapons, such as pikes, maces, and swords, there was little opportunity to concentrate attackers.[6] One soldier might face ten on the field of battle, but the nature of hand-to-hand combat meant that he was unlikely to face all of his opponents at once. There simply wasn't space enough for them all to approach him at the same time and, if they did, the swinging weapons of adjacent soldiers would get in the way. The reality was that most of the time, such engagements occurred in succession. Our lone warrior would face each of his opponents in turn, one after the other.

Armies of this age lined up against each other and clashed man-to-man in a long string of individual encounters. Reinforcements lurked in the wings, since there was no room for them on the line, and soldiers stepped up as needed to fill the gap when the man before them fell. At Agincourt (1415 CE), for example, 1,500 English infantry, knights, and men-at-arms suited in armor and carrying pikes and swords, faced off against 8,000 Frenchmen.[7] The armies stood shoulder to shoulder in parallel lines that crossed the full extent of the meadow, and reinforcements pressed in rows behind their respective front lines. The French outnumbered the English by nearly five to one, but this only meant they were stacked twenty deep instead of four. The actual battle was fought man-to-man.

In conflicts such as this, Lanchester realized, victory hinged both on the number and effectiveness of soldiers. Specifically, the loss inflicted equaled the number of soldiers times the effectiveness of each soldier. Because the fights were face-to-face, the strength and training of each soldier, as well as the quality and often the size of his weapons, mattered a great deal. The best-armed soldier was the most likely to win in each of the individual engagements, causing the better quality army to lose forces at a slower rate. Soldier numbers mattered, too, of course, since a numerical advantage permitted an army to sustain prolonged engagements by cycling or replacing soldiers on the line, but the fighting ability of individual soldiers could tip the balance.

Lanchester's models describing ancient warfare are called his "linear laws," because the equations used to determine the rate of force loss are linear. He designed these first types of models so that he could contrast them with his second type, the models he was convinced mattered in the modern age. Lanchester recognized that long-range weapons would alleviate the "available space" constraint of hand-to-hand combat. Guns could be fired from a distance, and this meant that multiple soldiers could concentrate all their fire on the same target. If one army was larger than the other, the extras didn't have to sit idly by, waiting in the wings. Their effectiveness could be brought to full force immediately because they could all fire at the enemy at once. Now, when he calculated the rate of attrition of opposing armies, he found that the effects of force strength—numbers of soldiers—mattered a lot more than they had before. The rate of loss of an army equaled the effectiveness of each soldier times the number of soldiers *squared* (these models, not surprisingly, are called his "square laws").[8] To put this difference into perspective, if one army in ancient times was five times the size of another, as the French were at Agincourt, it would have been five times as powerful. Now, because of the multiplicative effects of troop number, this same army would be twenty-five times as powerful as its opponent.[9]

The genius of Lanchester's models was his realization that the capacity to concentrate fire—to have multiple soldiers confronting a rival simultaneously—utterly changed the formula for victory. Based on his models, military strategists quickly recognized that investing substantially in the training and arming of individuals was not cost-effective, because the fighting effectiveness of each man was less likely to determine the outcome of battle than the numbers of men brought to the fight. Obviously, some training and equipment were needed, since soldiers armed more poorly than their opponents would be less effective. But given the choice of allocating resources to weapons and training, on the one hand, or to adding more soldiers on the other, the winning strategy was clear. Add more soldiers.

Since Lanchester's pioneering work, his square law of combat has been applied to post hoc analyses of thousands of battles, ranging from Ardennes to Iwo Jima, and it has formed the foundation and inspiration for countless models of military force allocation, military strategy, and military spending.[10] Lessons from these original models, such as the near-sacred maxim of never dividing one's forces on the field of battle, still permeate the military mindset today.[11]

• • •

Lanchester designed his square law for modern warfare and, without question, the vast majority of interest has focused on that category of equations. But it's his linear law that is most relevant to the evolution of extreme weapons. By contrasting ancient and modern warfare, Lanchester helped define the circumstances in which large weapons will, and will not, be cost-effective. When opponents can concentrate their fire—gang up on an opponent simultaneously—investing in large weapons is probably a mistake. On the other hand, when soldiers duel each other at close range, face-to-face and one-on-one, then the better fighter is likely to win. Because fighting ability often depends on the size of a weapon, duels can lead to situations where bigger and bigger weapons prevail.[12]

Duels matter for animal weapons, too, for essentially the same reason. Animal fights occur in all sorts of crazy places, from steep rocky cliffs to canopies of tropical trees and thermal vents on the ocean floor, and the details of these contests vary no less spectacularly. Some of these battles lend themselves to the use of weapons while others do not.

With the exception of social insects such as ants and termites, most animals do not fight with armies.[13] Males fighting for opportunities to mate do so as individuals. They fight for themselves. But this does not mean that all males confront each other one-on-one. In fact, many encounters are far wilder than this, involving chaotic scrambles as piles of rivals all attack at once. Just as Lanchester contrasted ancient and modern soldiers so, too, we can contrast animal battles that occur as duels with those that occur as scrambles.

When males attack each other face-to-face, the battles tend to be stereotyped and repeatable—weapons lock, opponents strain against each other, and push or pull or twist, depending on the species. Battles become reliable tests of relative strength, and the best quality male generally wins. When rivals scramble together in chaotic tumbles, the outcome of fights is less predictable and the value of weapons diminishes.

Male cicada-killer wasps pounce on each other in vicious midair tangles, twisting and writhing and biting, often crashing to the ground in the process.[14] Fights occur in the open, above hard-packed soil containing preadult females that have yet to emerge, and it's not uncommon to encounter three or four males locked in combat at the same time. Females complete their development underground, metamorphosing from larva to pupa to adult in the relative safety of the soil. They lie in clusters, climbing up to the surface unmated and reproductively receptive. Males can smell where these females lurk, and dozens may arrive to fight for possession of the prized real estate hiding them. Victorious males grab females and mate with them as soon as they appear—sometimes even helping them along by digging them out of the ground. Cicada-killer wasps experience intense male competition for localized resources (buried females), but they lack elaborate weapons.[15]

Horseshoe crabs swim ashore under full and new moon high tides in fantastic swarms. Hundreds of thousands of individuals clamber out of the sea to mate in the moonlight, blanketing beaches in a deep carpet of white foamy sperm. Here, as with so many animals, reproductive females are few and far between. By the time a gravid female swims ashore, she will already have a male latched onto her back. But he'll have to hold on tightly to have a shot at fertilizing her eggs, since males challenge him from all sides.[16] It's not uncommon for females to have four or five males piled onto their backs, pushing and scrambling and jockeying for position. Yet these males, too, lack significant weapons. Both cicada-killer wasps and horseshoe crabs face intense competition for females who are localized and defensible, the first two conditions for an arms race. But fights unfold as scrambles rather than duels. They lack the condition of Lanchester's linear law, and large weapons are not cost-effective.

Although the concept of "fair" is a human construct, it reflects, in a way, the predictability and consistency of outcome. In a fair fight, the best fighter should win (any upset to this outcome would be perceived as cheating), and the fairest fights are always duels. Since the beginning of recorded history, in military traditions ranging from Homer's Greeks to medieval knights, samurai, and gunslingers in the American Wild West, the only form of confrontation that has ever been acceptable for establishing honor, status, or rank was the duel.[17]

In animal duels, too, the best fighters usually win, whereas in scrambles they may not. Head-on encounters are relatively predictable and straightforward, without surprises. In these fights it is harder for a poor quality male to usurp a bigger, better male. As with ancient warriors, strength, stamina, and weapon size prevail.

• • •

All else being equal, we expect species where males fight each other one-on-one to be more likely to evolve extreme weapons than species where males fight less predictably. But what habitats are conducive to animal duels? It turns out that many of the same ecological situations that cause resources to be economically defensible also align, or otherwise constrain, male contests so that they tend to occur as duels. By acting as choke points, these special situations become cauldrons for the evolution of extreme weapons.

Burrows are probably the most widespread ecological situation leading to the evolution of extreme weapons. They are localized and readily defendable. Males can guard the entrances to tunnels where females reside, and in so doing, block rival males from approaching the females. But tunnels also physically restrict access in a way that aligns the interactions of opponents. A rival male dung beetle has to enter the tunnel before he can challenge the guarding male. Ten males couldn't attack at once even if they wanted to, because there isn't space for more than one rival to enter at a time. The restricted confines of tunnels align battles so that they necessarily occur as a series of successive duels. Ball-rolling species, on the other hand, face no such restricted access. Males can challenge from all angles at once, and very often battles among ball rollers entail chaotic scrambles between three or four males. In dung beetles, species that fight one-on-one often have elaborate horns; species that fight in chaotic scrambles do not.

Shrimp and crabs with huge claws battle over burrows.[18] Wasps with long tusks fight over the tubular entrances to mud-pot nests—burrows—that they cement onto the undersides of leaves.[19] Many species of rhinoceros beetles fight over burrows, either tunnels in the soil or hollowed-out stems of plants such as sugarcane.[20] There are even a few species of unusual Asian frogs that fight over burrows, and these same species are unique among frogs in also bearing male fangs and spurs.[21] And there is evidence that an extinct and giant species of horned gopher fought over burrows.[22] Although by no means a guarantee of weapon evolution, burrowing behavior provides two of the three critical prerequisites for an arms race, and this appears to have tipped the balance in species after species.

Branches work the same way. In essence, they are "inverse tunnels" since a branch, like a tunnel, is a linear substrate that can be blocked and along which a rival must pass. Like trolls guarding bridges in fairy tales, males can plant themselves as gatekeepers. To get to a female on the other side, a rival must challenge the guarding male, and because the branch is long and narrow, only one male can approach and fight at a time. Animals as diverse as rhinoceros beetles,[23] leaf-footed bugs,[24] and horned chameleons[25] defend branches, blocking the passage of rival males and gaining access to females in the process. Males in many of these species invest in elaborate weapons.

Even exposed locations can align male interactions if the critical resource is stationary and small enough for a male to stand over it and guard. Males plant themselves atop the resource and swivel as needed to face each attacker, like an ice fisherman protecting the hole he's drilled in a lake. Stag beetles battle over sap oozing from nicks on the sides of standing trees, much like the harlequin beetles. Males grasp each other head-on with their mandibles and strain, lifting with their bodies and legs, trying to rip their opponents free from the tree trunk and fling them to the ground below. Females visit these sap sites to feed before they fly off to lay their eggs, and victorious males mate with them while they are feeding.[26]

New Guinean antlered flies fight over tiny holes in the bark of fallen trees. Females must get through the bark to lay their eggs, and they can do this only using an existing hole (the flies are not robust enough to drill holes of their own). Males capitalize on this situation and stand guard over the holes, diligently moistening and marking them to attract females and fighting with all trespasser males who come by.[27] Because males can stand over the resource, they can turn to face each rival head-on.

In all of these examples, males compete for females, and they control access to females by fighting to guard something that is economically defensible. And, in each of these species, the very feature of the habitat that rendered resources defensible structured male contests so that they occurred as duels, rather than scrambles.

• • •

Eyes in male stalk-eyed flies sprout like lollipops from the sides of their heads, giving these flies an uncanny resemblance to miniature sets of barbells. Males in some species have absurdly long eyestalks, while males in very closely related species do not. As with dung beetles, we can explain this variation by looking closely at the natural history of each stalk-eyed species.

Ingrid de la Motte and Dietrich Burkhardt studied natural populations of five species with long eyestalks, and several additional species without them.[28] What they found exactly fits our prediction.

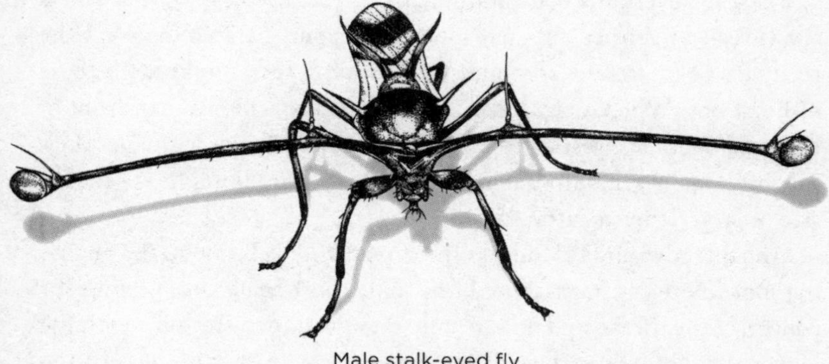

Male stalk-eyed fly

Species such as *Teleopsis whitei* and *T. dalmanni*, each with males with huge eyestalks, spend their daytimes walking along the ground or on low vegetation near forest streams, feeding on fungi, molds, and yeast from decaying leaf litter or dead animals. During the day they forage alone, and they are aggressive toward other flies—male or female—who approach.

Each night, these same flies cluster together to roost in dense aggregations on dangling perches in the forest. In sheltered alcoves beneath the undercut banks of small streams, tiny rootlets hang down like threads. Some rootlets are longer than others, and long threads can hold more flies than small threads. Females often crowd onto these threads in groups of as many as twenty or thirty, forming a linear hanging harem.

From a male fly's perspective, rootlets are critical resources routinely used by females. For the few males able to do it, guarding a thread means securing disproportionate access to females, which, given the numbers of females per thread, translates into a huge reproductive benefit. A number of biologists have now hiked up tropical streams in Africa and Asia to study these flies at their nighttime roosts. Two major research groups, one headed by Gerald Wilkinson and including John Swallow and Patrick Lorch,[29] and the other including Andrew Pomiankowski, Kevin Fowler, and Sam Cotton,[30] have all spent long nights with headlamps watching males fight to guard their hanging territories.

The dominant male perches near the top of the thread, sometimes swaying his eyestalks back and forth in a rocking motion that twists the thread in gentle, sinuous ripples. Flies may be able to assess the relative size of a male from afar by observing the amplitude of these undulations. When a rival male approaches, he will hover in front of the resident male, eyestalk to eyestalk. If the newcomer is smaller than the resident, he generally departs without incident. But if he is equal in size or larger, then a battle ensues.

The intruding male lands on the thread and walks up to the guarding male. Forelegs outstretched, the males butt heads and grapple for control of the thread and, in virtually every instance, the male with the longer eyestalks wins. The winner then mates with each of the roosting

Male stalkie guarding a harem

females during the night. For these species, the benefits of territory defense appear to vastly outweigh the costs of producing and bearing a weapon, even a truly enormous and awkward one.

Remarkably, when people studied related flies that lacked large eyestalks, they found that the single biggest difference in their behavior was that these flies did not roost communally at night. Flies like *Teleopsis quinqueguttata*, which have only rudimentary eyestalks in males and females, never grouped into defensible clusters. Like their relatives, they fed in isolation during the daytime on molds and fungi. But at night they roosted alone dispersed in vegetation. All mating occurred in the daytime, in happenstance and brief encounters between the sexes. No duels, no weapons.

• • •

When Lanchester modeled duels in his simulated ancient battles, he pictured soldiers grappling with other soldiers, but his logic works just as well for confrontations between larger entities. Ships attack other ships, fighter planes attack fighter planes, and nation-states attack rival nation-states. For these confrontations, too, the nature of the interaction matters, and opponents lining up one-on-one can spark an arms race. For example, for almost 1,500 years oared galleys churned through Mediterranean waters as Egyptians, Phoenicians, Carthaginians, and Greeks all battled for supremacy.[31] For most of this period (1800–750 BCE) ship design remained largely unchanged. Fleets of long, canoe-like ships shuttled soldiers back and forth to battle, powered by sail when the winds cooperated and at all other times by sweat and muscle, as rows of oarsmen spaced along the length of each side pulled long oars in unison to propel the vessel. But somewhere around 750–700 BCE everything changed. A new weapon was added to the galleys: the battering ram.

Cast from the finest quality bronze forged in the best kilns of the day, rams permitted well-manned ships to smash into other ships, shattering their hulls and sinking them along with their crews. Because of rams, naval ships could be more than simple vessels. They could be weapons. Ships suddenly acted as units—individuals—and they confronted each other up close and one-on-one. Maritime battles began to resemble the clashes of ancient infantry, with ships lined up abreast in long rows, crashing into opposing lines of ships from another fleet.[32] The battering ram caused naval warfare to fulfill the conditions of Lanchester's linear law: ships fought other ships in close range duels. From this point onward, bigger was better and navies with the largest ships won.[33]

What unfolded was one of the grandest naval arms races of all time, as shipwrights struggled to add speed and power to their vessels, and each new innovation from one side was instantly copied and then bested by the other. Early galleys such as the penteconter had roughly 25 oarsmen per side, and the first attempts to add speed and power involved lengthening the hull to add more oars. But ships rapidly reached a maximum at roughly 130 feet, beyond which the hulls buckled in rough seas.[34] By around 600 BCE modifications to the hull permitted greater height, and a second tier of rowers was added above the first, doubling the power. These new ships could pack their power into shorter hulls, which were stronger and more maneuverable than longer ships.[35] The bireme had a wooden hull just 80 feet long and 10 feet wide, but it now housed 50 oarsmen per side, for a total of 100 oars. Triremes soon followed, with the addition of still greater height and a third tier of oars. Triremes grew to 130 feet long and 20 feet wide, propelled by 180 oars. The trireme reached a maximum, however, since increases in ship length meant buckling, and increases in ship height (for additional rows of oars) meant tipping.

For almost two hundred years the trireme prevailed as the dominant ship of the line for naval fleets, until yet another modification to the hull opened the way for even greater increases in size. Thus far, each galleon had housed just one man per oar. Much like the trunk of a centipede, which has a string of repeated units sprouting a leg from each side, the ancient galley comprised a string of sections, with oars extending from port and starboard. Penteconters were called "ones" because one man pulled the one oar projecting from each side of every segment. Biremes were called "twos" because two men sat on each side of every segment, one above the other, each pulling one of the two oars. Triremes were "threes" for the same reason. But by the fourth century BCE shipwrights had discovered that they could add power,

The first naval arms race, before and after: a penteconter contrasted with Ptolemy's "forty."

and with it speed, by cramming additional men into these confined spaces.[36]

"Fives" were created by adding a second oarsman to two of the three oars in a trireme. Fives still had 90 oars per side (in three rows of 30), but these oars were now powered by 300 men instead of 180. Bigger really was better, and fives bested threes in a fair fight. By 387 BCE "sixes" were added to the line, and within ten years ships had increased through "sevens," "eights," and then "nines" (nines had three men abreast at each of three oars). By 315 BCE there was a "ten," and by 301 BCE fleets had experimented with "elevens," "thirteens," a "fifteen," and a "sixteen." By this point ships were getting unwieldy, however, and most ships above tens were considered clumsy and slow, though they made impressive displays of power. The culmination of the race was a behemoth contracted by Ptolemy V, the "forty." This beast of a ship likely consisted of two parallel hulls bridged catamaran-style by an upper deck (more rowers could fit this way, since each hull had files of oarsmen along each side). More than 420 feet long and powered by 4,000 oarsmen, Ptolemy's forty was the largest vessel ever built in antiquity.[37] Not surprisingly, it was so extreme that it was nautically worthless.

• • •

Together with competition and economic defensibility, duels tip the evolutionary balance in favor of extreme weapons. This simple insight has terrific explanatory power because it provides a general rule for why particular animal species have extreme weapons, and why their close relatives may not. Like uncovering the key to an ancient encrypted cipher, we can now make sense of so much animal diversity. Specifically, we can look to the histories of groups of species to see when behaviors changed in ways that caused the arms race ingredients to fall into place. This can help explain why isolated species stand out from the rest, and it can also explain why whole groups of related species appear to enter into arms races together. Sometimes, one or more of the arms race ingredients are part of the inherited physiology of a group of species, and this can predispose them all to arms races.

As groups of organisms diversify through time, spawning new lineages that turn into new species, they carry with them legacies of their past. Mammalian carnivores all inherit the same basic types of teeth (canines, premolars, molars, and so forth) because all carnivore species descend from a common ancestor that had that set of teeth. There are just over fifty species of field mouse (*Peromyscus*) scattered around the world, and they all use the same enzymes to infuse pigments into their fur because their ancestors, a mouse species that lived roughly ten million years ago, used these enzymes to pattern its fur.

Sometimes the characteristics inherited by a group of species (species that trace their ancestry back to a shared common ancestor are called a "clade") cause many members of the clade to all experience similarly strong selection for large weapons. For example, female African elephants invest impressive amounts of time, energy, and nutrition into their offspring during multiyear bouts of pregnancy and nursing. These are major features of their physiology and behavior, and this extreme form of investment was likely shared by many other species within the elephant clade. In fact, we know from specimens hacked out of glacial ice and pulled from anoxic tar pits that female woolly mammoths, Columbian mammoths, and mastodons also nurtured their young through a prolonged pregnancy. Paleontologists can tell from the shapes of fossil pelvic bones that all of the extinct elephant relatives did as well.[38] This means that the crucial backdrop for intense male-male competition in African elephants, a highly skewed ratio of available males to available females, almost certainly existed in all of the other species in this clade, too. It's probably no coincidence that so many relatives of elephants embarked on paths of rapid and extreme weapon evolution.

This is why we often find not just a single species with huge weapons but entire clades packed with species after species all armed to the teeth. Inherited characteristics like asymmetrical parenting tip the balance in favor of arms races for all descendant members of the clade. All that is needed is for the remaining two pieces to fall into place. If many of these species also share another one of the ingredients, say an inclination for using habitats such as burrows that result in defensible resources or choke points, the balance is tipped still further. The result can be explosions of animal diversity, as species after species within these clades launch onto trajectories of rapid weapon evolution.

There are more than one thousand species of stag beetles worldwide, and almost all of them have extreme male weapons.[39] Stag beetles form a separate branch on the beetle evolutionary "tree" from the dung beetles and rhinoceros beetles. Instead of horns, they produce a pair of toothed mandibles so large that in some cases they are longer than the male himself. Stag beetles descend from an ancestor that likely experienced strong sexual selection and defended sap sites on the sides of trees, as almost all living stag beetles still do today. This shared suite of characteristics appears to have predisposed stag beetle populations to escalated evolution of large male weapons. Extreme mandible sizes evolved at least twice early in the history of this group.[40] These large-mandibled species subsequently radiated into literally hundreds of daughter species who continue to this day to experience intense sexual selection, with males fighting in duels over localized oozes of sap.

The same insight helps explain extreme weapons in flies. The vinegar flies, or Drosophilidae, contain more than three thousand species, and the vast majority of them lack extravagant male weapons. However, at least eleven different times in the history of this group, males evolved extensions to their heads that functioned in fights with rival males. (Both the stalk-eyed flies and antlered flies are clades of "big-headed" flies within the larger vinegar fly family; that is, they make up two of the eleven evolutionary origins of extreme male weapons.) When David Grimaldi, a curator at the American Museum of Natural History, looked closely at the biology of the flies, he concluded that all of these exceptional species stood out from the pack in the same three ways. Unlike other Drosophilid species, males with big weapons all displayed unusual levels of aggressive competition, guarded localized resources, and faced each other head-on in fights that were either described as "head butting" or "jousting."[41]

After the extinction of the dinosaurs (about sixty-five million years ago), mammals dominated terrestrial landscapes, and the ungulates, in particular, thrived. These hoofed plant eaters diversified in droves, as group after group expanded, diverged into clades of related lineages, and eventually disappeared. Punctuating this history was an exuberance of clades bursting with extreme weapons.[42]

Arsinothere and *Synthetoceras*, early ungulates with unusual weapons

Brontotheres began no larger than modern coyotes, but they rapidly evolved into giants standing eight feet high at the shoulder and weighing twenty thousand pounds. Early brontotheres had no weapons; later species bore broad, flat, bony plates on their noses that could extend more than two feet. Rhinoceroses began small, too—dog-sized—and hornless, diversifying later into enormous animals weighing up to thirty thousand pounds and wielding dramatic weapons. The woolly rhinoceros, for example, had a horn more than six feet long. At their apex, the rhinoceroses comprised more than fifty species worldwide, but most have disappeared, and only four species still survive today.

Around the same time, the snouted ungulates diversified. Beginning with small, unarmed, early elephants, they radiated into more than 150 species with weapons like the "shovel-tusks," with three-foot lower incisors jutting forward as flat blades from the lower jaw; the "hoe tusks," with downward-curving tusks curling beneath the lower jaw; the "upper tusks," as in mastodons and modern elephants; and even the "four tusks," with two upper and two lower tusks.

But the ungulates were only just getting started. A clade of pigs diversified into species with a unicorn-like head horn and species with long, curling tusks. An offshoot of the camels exploded into crazy armed forms such as *Synthetoceras*, which sprouted a pair of horns from the back of the head along with a huge, forked horn protruding up from the snout, to *Kyptoceras*, which had two long horns curving forward from the back of the head and a sideways pair of horns arcing like pincers above the nose. A clade of pronghorn antelope gave rise to dozens of species with elaborate horns, and the giraffes radiated into at least ten species with bizarre and diverse horns. Last, but not least, the deer began as tiny fanged animals like the modern-day Chinese water deer, but rapidly diversified into almost a hundred species with bony antlers famous for their size and complexity.

What these broad patterns reveal is a simple and surprisingly universal rule: after the final ingredient for arms races falls into place, entire clades of descendant species can all experience rapid evolution of extreme male weapons. Mastodons and flies could not be more different from each other. They lived at different times, in different habitats, and fed on different foods. One was more than 120 million times

Weapon diversity in deer

the size of the other. One had enlarged teeth, and the other had chitin protrusions from the forehead. Yet, the same three ingredients triggered evolution of extreme weapons in both cases. So it is with wasps, beetles, crabs, earwigs, elephants, and antelope. Despite extraordinary differences among these species, arms races are arms races, and the circumstances leading to big weapons are the same.

PART III

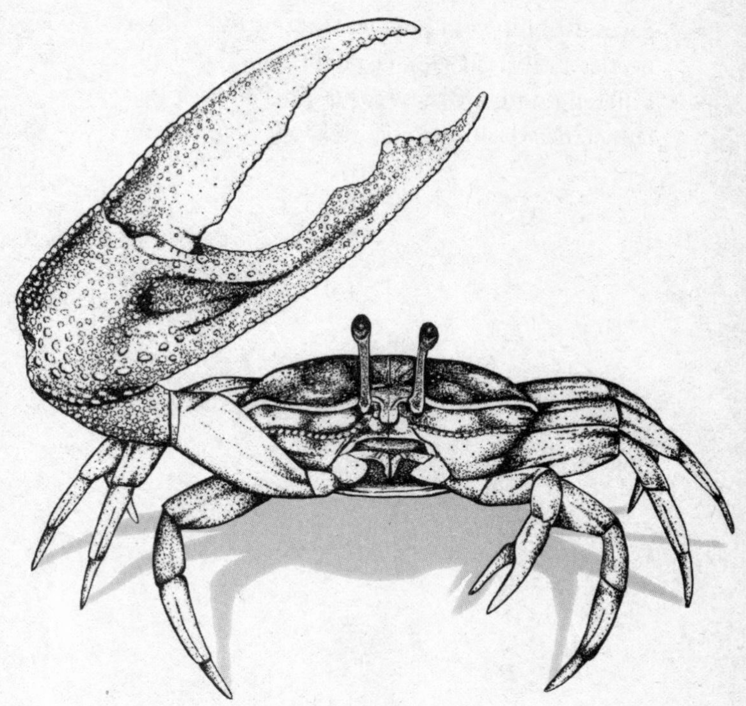

RUNNING ITS COURSE

After a race is triggered, weapons begin getting really big, and several things happen along the way. Appreciating the stages of a race reveals striking similarities across species—a surprising suite of characteristics shared by all of the most extreme weapons, including our own.

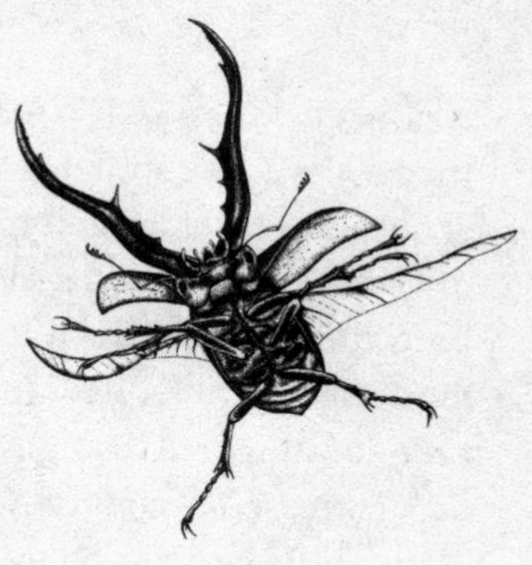

7. Costs

Far below, the waters of Gatun Lake lay dotted with green and red as channel markers of the Panama Canal blinked in the moonlight. It was five a.m. and I was sitting in bed staring out at the shadowy branches of a rainforest canopy. My room was on the upstairs floor of a small laboratory surrounded by lush tropical forest, at the top of a very steep slope. The lab was a simple wooden structure with four dorm rooms in a row. Mine was at the end, and three of its four walls were encased only in screening. The moist wind and spray of rain passed straight through the room, covering my face and the sheets of my bed. The forest teemed with whines and chucks of túngara frogs, eerie trills of cane toads, and, of course, the ever-present drip of rainwater from the leaves and eaves overhead.

I'd awakened before dawn that morning as I did every day, to listen for a particular sound that would lead me to the beetles I'd come to the island to study. It was August 1991, well into the rainy season, and I was a doctoral student there for a stint on Barro Colorado Island to conduct field research into the function of horns in beetles. Although the beetles I was studying that year were abundant, they were tiny—about the size of an eraser on a new pencil—and they could be very difficult to locate. The trick to finding them was to find their food, which, unfortunately, happened to be howler monkey dung. So my task each morning was to locate a troop of howlers before they left their nighttime roosting tree. Dung materialized fairly predictably with monkey departure, so if I could find that spot fast enough, I could catch these elusive little beetles as they arrived.

That morning it didn't take long for me to hear what I was waiting for. Almost like clockwork, just before the first light of dawn, came the throaty roar of the howlers announcing their location to rivals in adjacent territories. When the monkeys were nearby, their howls were painfully loud. But those were good mornings because the monkeys and beetles could be located quickly. On other mornings the dawn roar was just barely discernible above the din of the forest night, and I might have a mile or more to trek before I found them. My routine was simple: take a compass bearing and then go back to sleep. An hour later, when it was actually starting to get light inside the forest, I'd head into the understory mist to track down that day's troop.

Sun streaked through cracks in the canopy, trapping clouds of steam in angled slabs of light as I jogged toward the now-silent monkeys, guided only by my compass. I pushed forward through the branches, stepping over roots, listening for movement in the canopy above. A branch rattled, and I'd found them. Ten faces glared down at me, coal-black shadows against the leaves; one of them hurled a stick.

Once I found the monkeys it was a quick task to find the dung, and it never took long for the insects to arrive. Big, shiny, metallic beetles landed with a plunk and clambered over twigs and leaves. Tiny yellow and brown beetles perched on leaves with their antennae outstretched, reaching for scent. Within minutes they were everywhere, undulating back and forth in tiny sweeps as they maneuvered clumsily into the pieces of fallen dung. Soon still other hovering, pea-sized beetles zigzagged over to the dung fragments. Swirls of flies landed first on surrounding leaves before hopping onto the dung as well. In less than an hour the forest teemed with insects that had all converged on the dung to find food and mates.

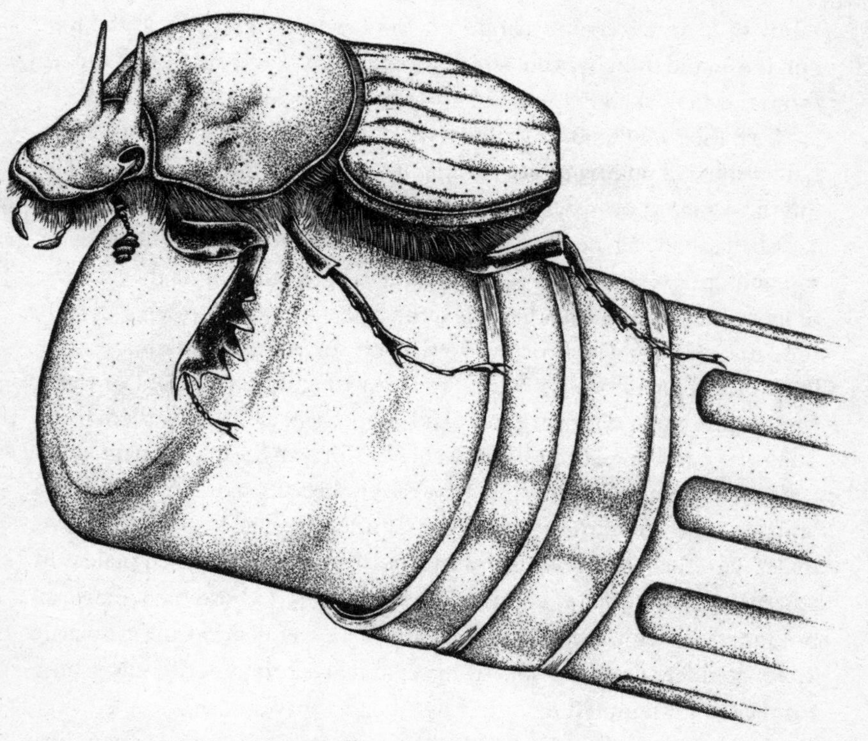

Onthophagus acuminatus, my Panama beetle

The species I was after that day has no common name and is rarely noticed by anyone other than entomologists.[1] The largest males wield a pair of horns, two cylindrical spikes rising side by side in a line between the eyes. (Smaller males, incidentally, do not have horns; they develop with only nubbins where the weapons would be.) My objective that year was to watch these horns evolve. I planned to accomplish this by applying selection pressure to the beetles myself in the dusty shed that served as my laboratory. I had purchased in bulk a ten-foot-tall bag of unlabeled plastic shampoo bottles from a company in Panama City. In the field station wood shop I cut the top off of each bottle with a band saw, so that they formed cylindrical tubes twelve inches deep and three inches in diameter. Almost a thousand of these tubes lined the counters of my little screened-in lab, each one filled with ten inches of packed, moist, sandy soil. An ice cream scoop's worth of monkey dung sat at the top, and the whole thing was enclosed by screen mesh and a rubber band. A thousand furnished homes, if you happen to be a dung beetle.

Each tube would house a single pair of beetles, going about their nuptial business of pulling pieces of dung into a tunnel and fashioning them into finger-sized compacted sausages called "brood balls." On the tip of each ball sat an egg, perched on a tiny stalk and encased in a thin shell of soil and dung. Once the egg hatched, the larva would spend the entirety of its development inside the little dung ball, eating and growing in solitude until it was ready to crawl to the soil surface as an adult a month later. Each pair of beetles could provision roughly six to eight of these egg-containing sausages in a week and I kept them at it, giving them fresh dung every few days, until I had twenty or thirty offspring per pair.

I started with one hundred wild-caught beetles, half males and half females. I measured the males under a microscope, and selected the five males with the longest horns to serve as breeders. Each chosen male was paired with two different females in succession. (I chose the females at random from within the lab population, since females don't have horns.) The mated females were housed in separate shampoo bottle tubes, and from each I attempted to collect between twenty and thirty offspring. Ten reproductive females times thirty offspring each yields roughly three hundred progeny per generation. Males of this generation would again be measured and, as before, the five individual males with the longest horns relative to their body size would be selected as breeders. Each male would be paired with two females apiece, and their offspring would comprise the third generation of the experiment, and so on.

The logic of artificial selection experiments is pretty straightforward. In my case, my breeding program selected for increased horn length in generation after generation of the population. The empirical question then concerned whether or not the population evolved in response to this selection. Did the horns of males get longer with each successive generation?

Treatments in a scientific experiment need to be replicated, however, to minimize the possibility of arriving at results by chance. Populations shift gradually from generation to generation simply as a result of serendipity. Picture a large jar full of fifty different colors of jelly

NOCYBE, WIKIMEDIA COMMONS

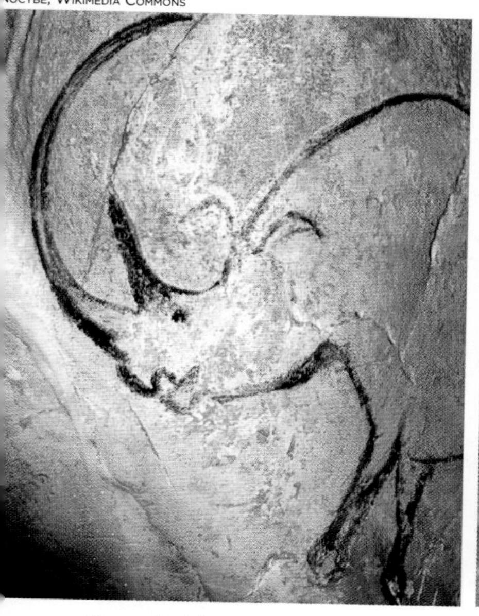

HTO, WIKIMEDIA COMMONS

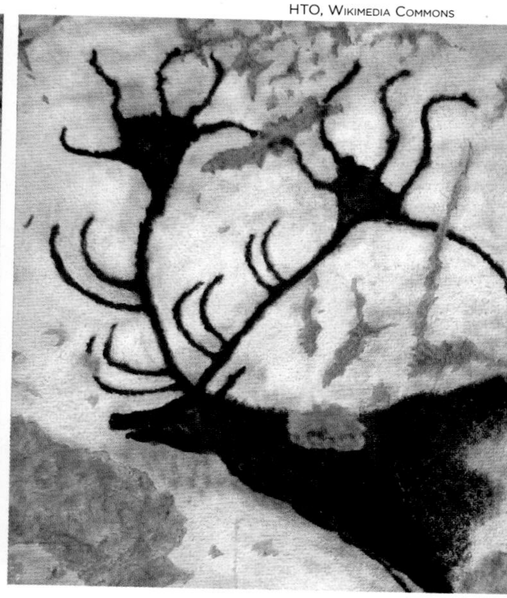

The earliest human paintings—some more than 30,000 years old—portray animals with extreme weapons. Wooly rhinoceros from Chauvet Cave (*left*) and Irish elk from Lascaux Cave (*right*).

Elk pay a steep price for their weapons. Bulls double their daily energetic needs and leach essential minerals away from other bones in order to grow antlers.

BY AUTHOR

Many animals, such as this katydid, rely on camouflage to avoid being eaten by predators. Leaf-mimic katydids even rock back and forth when they walk, resembling a leaf fluttering in the wind.

Nicholas Benroth, Wikimedia Commons

Special Forces snipers also rely on camouflage, choosing from among a selection of specialized suits the one most likely to match the particular background of each mission. Exquisite background matching is not cost effective for most other situations, and the majority of soldiers wear uniforms that provide only moderate camouflage—a universal color pattern designed to work reasonably well against a range of different backgrounds.

Geoff Gallice, Wikimedia Commons

Animals employ diverse defensive weapons, including spines and armor.

BY AUTHOR

Cats are agile predators whose weapons are moderate in size, reflecting a balance between selection for killing large prey and selection for agility and speed. *Top*: Mountain lions behind the author's house. *Bottom*: Canada lynx.

KEITH WILLIAMS, WIKIMEDIA COMMONS

Dumas, Wikimedia Commons

Weapons in ambush predators experience very different selection from weapons in other predators. Instead of running, swimming, or flying fast to catch prey, ambush predators snatch them with a quick strike from claspers. As a result, many have large weapons. *Top*: Mantisfly. *Bottom*: Peacock mantis shrimp.

Silke Baron, Wikimedia Commons

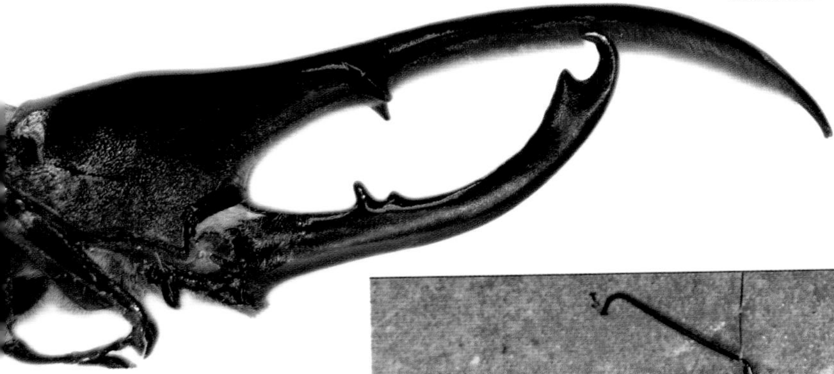

By author

By author

The most extreme animal weapons are wielded by males, and used in battles for access to females. *Top*: Rhinoceros beetle horns. *Bottom*: Harlequin beetle forelegs.

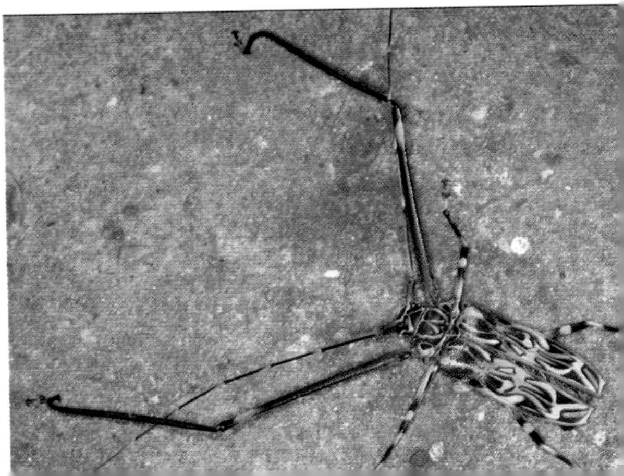

Diversity in dung beetle horns. All of these species are "tunnelers," who use their weapons in fights for burrows containing females.

Udo Schmidt, Wikimedia Commons

Udo Schmidt, Wikimedia Commons

by author

by author

by author

by author

JLeber, Wikimedia Commons

When males fight in unrestricted places, such as in the air, or in chaotic scrambles, agility takes precedence over big weapons. *Top*: Cicada killer wasps. *Right*: Horseshoe crabs.

Hayden, Wikimedia Commons

Gerald Wilkinson

Males in many animals, such as these flies, fight over restricted, economically defensible resources visited by females. In these situations, big weapons are often beneficial. *Left*: Stalk-eyed flies defend hanging rootlets on which females roost. *Bottom*: Antlered flies defend egg-laying holes in the bark of fallen trees.

Densey Clyne

Duels are an essential ingredient of an arms race. When rival males face off one-on-one, selection often favors the male with the bigger weapons, driving evolution of extreme weapon sizes. *Top*: Stag beetle males. *Bottom*: White-tailed deer.

Donald M. Jones

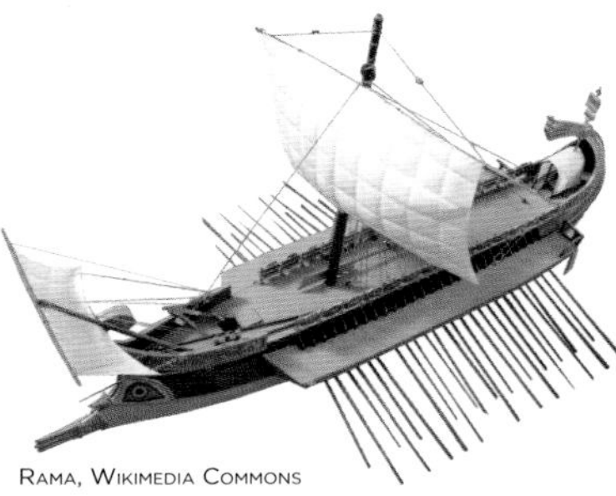

Rama, Wikimedia Commons

John Christian Schetky, *The Furious Action between H.M.S. Mars and the French '74 Hercule off Brest on 21st April 1798*, Wikimedia Commons

Vehicles also often fight in one-on-one duels and, as in animals, this can trigger an arms race. *Top*: The addition of a battering ram precipitated explosive growth in ship size in oared galleys of the ancient Mediterranean, settling in the end on the "five." *Right*: Side-mounted cannons had the same effect on sailing galleons.

Mark Bridger, Shutterstock

Erwin and Peggy Bauer, Wikimedia Commons

Fallow deer and moose pay exorbitantly for their weapons, and for the stamina and energy needed to win fights during the rut, suffering gashes, infections, and depleted energy reserves. *Top*: Three-quarters of fallow deer bucks die without ever succeeding in defending territories, and 90 percent fail to mate even once in their lifetime. *Right*: Bull moose double their daily energetic demands while growing antlers, and a third die from injuries sustained during combat.

JOHAN SWANEPOEL, SHUTTERSTOCK

Battling for access to females is dangerous, resulting frequently in serious wounds and, occasionally, death. *Top*: Fighting male oryx. *Bottom*: Male antelope died after their horns locked together in combat.

ERICK GREENE

Geoff Gallice, Wikimedia Commons

Top: Relative to their body size, male fiddler crabs wield the largest weapons of any animal. *Bottom*: Because fighting is dangerous, males assess each other beforehand, comparing weapons and pushing. Claws function as deterrents, and fights only escalate when males are evenly matched.

Andrea Westmoreland, Wikimedia Commons

THOMAS WHITCOMBE, *A CROWDED FLAGSHIP OF AN ADMIRAL OF THE BLUE PASSING MOUNT EDGECOMBE AS SHE CLOSES INTO PORT AT PLYMOUTH*, WIKIMEDIA COMMONS

Ships function as deterrents, too. *Top*: Capital "ships of the line" were the most expensive and powerful weapons of their day. Out of reach to all but the richest nations, these ships projected power to the farthest reaches of the globe. *Bottom*: Nimitz-class aircraft carriers function in this same capacity today.

OFFICIAL NAVY PAGE, U.S. NAVY, WIKIMEDIA COMMONS

BY AUTHOR

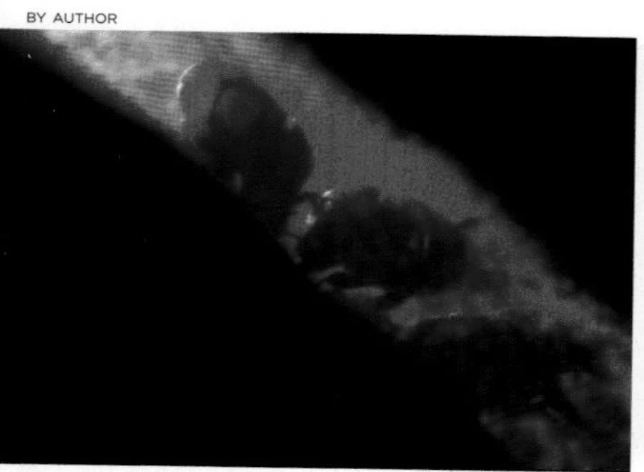

SMITHSONIAN TROPICAL RESEARCH INSTITUTE

Searching for sneaky males. *Left*: Male dung beetles fight inside a tunnel; the female is beneath them to the right. *Right*: Observing behavior inside tunnels is possible using "ant farms." Small, hornless males sometimes sneak into guarded tunnels below ground, digging side tunnels that intercept the main tunnels beneath guarding males.

Wikimedia Commons

Army ants overwhelm their prey with strength in numbers. Soldiers have big heads with strong jaws, but they still need to be mobile since they march along with columns of workers on raids.

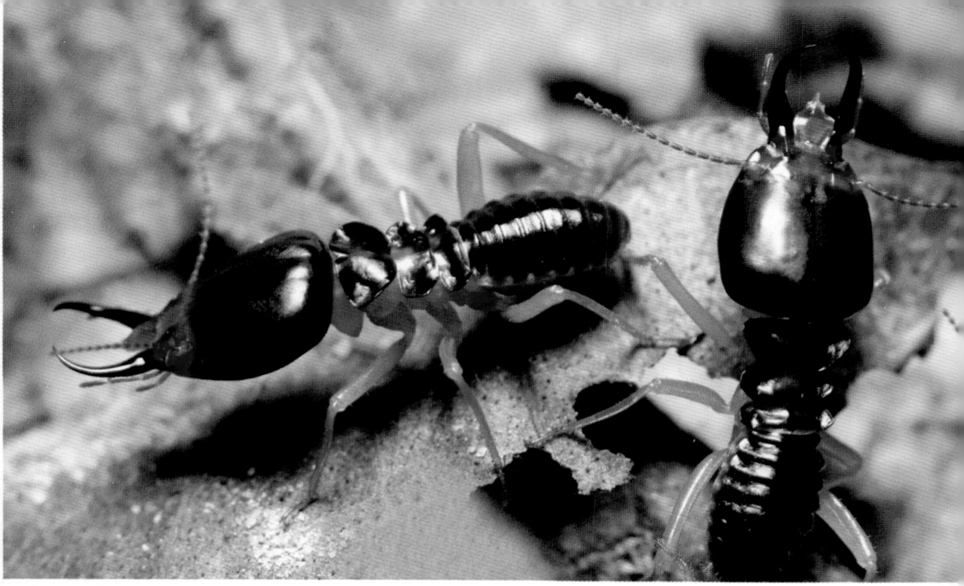

SYDEEN, SHUTTERSTOCK

Top: Termite soldiers have even bigger heads than army ant soldiers, in part because, unlike ants, they do not need to run long distances. Instead, they position themselves at the entrances to tunnels, biting anything that attempts to enter. *Bottom*: Termites rely on fortresses to defend themselves against attacks by army ants. By restricting access to a small number of tunnels, termites remove the numerical advantage of the army ants, forcing their soldiers to confront termite soldiers in duels the termites are likely to win.

ALI MUFTI, SHUTTERSTOCK

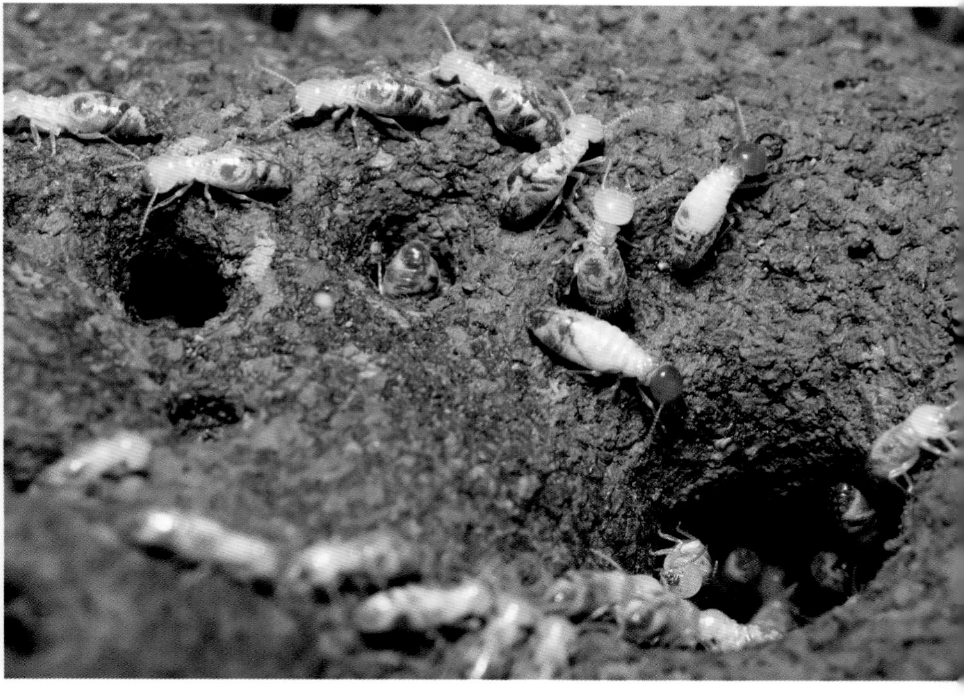

Castles rely on heavily guarded, tiny entrances, too. Gatehouses work like tunnels of termite nests, stripping away the numerical advantage of attacking armies.

Rachelle Burnside, Shutterstock

Fulviusbsas, Wikimedia Commons

Left: Fortification styles evolved in tandem with ever more effective artillery. Early structures relied on square-sided, protruding towers to provide flanking fire along the walls, but corners proved vulnerable to flying boulders. *Bottom*: Later castles had round towers to deflect artillery. Even when struck directly, these towers were less likely to shatter.

Antony McCallum, Wikimedia Commons

Gebruiker, Wikimedia Commons

Top: Cannons destroyed even the most magnificent castles, shattering towers, breaching walls, and collapsing the arms race, until a new style of fortress emerged. Star forts sit low to the ground and rely on extensive earthworks to absorb the impact of cannons, and angled, pointy walls to deflect cannonballs fired from any direction. *Bottom*: Exploding artillery spelled the end for even these forts. From WWII onward the safest place to hide has been in dispersed bunkers deep belowground, like Cheyenne Mountain, headquarters of NORAD during the Cold War.

U.S.A.F., Wikimedia Commons

Paulus Hector Mair, *De Arte Athletica II*, Wikimedia Commons

Medieval knights fought as individuals in duels with rival knights. *Top*: The most common battles were tournaments—structured spectacles with strict rules—that tended to favor the more heavily armed and better trained contestor. *Bottom*: Even in full battle, knights confronted other knights in hand-to-hand combat that often unfolded as duels.

John Gilbert, *The Morning at Agincourt*, Wikimedia Commons

beans, all mixed together. Reach in and scoop out a thousand and move them to a new jar. Chances are, you will happen to include most, if not all, of the fifty original types. Some might be a bit better represented in your scoopfuls than they had been before, but these changes in flavor frequency are likely to be minor. The jelly beans in the new jar should resemble the blend of flavors that were present in the old jar.

If, instead, you scooped only five jelly beans from the original jar and these then populated your new jar, they almost certainly would not be representative. Most of the original fifty flavors would be lost. If those five jelly beans were then replicated until the new jar was full, the total number of jelly beans might be similar to what was present before, but the blend of flavors would be drastically different. The jelly bean "population" would have evolved simply due to chance.

I was selecting only five individual males and ten females to found each generation. This is a small sample, and it meant that some shift in the traits of my population could have occurred by chance. If, at the end of my experiment, the males in my artificially selected population had longer horns than before, I couldn't rule out the possibility that this was a spurious result due to chance.

To confirm my findings, I would have to do the whole thing again. If you have not one, but two separate populations, both experiencing directional artificial selection for longer horn lengths, and they both result at the end in males with longer horns, it's a much more compelling result. Random changes are not likely to occur in the same direction both times. Even better, add still more populations selected in the opposite direction—males with the shortest horns selected as the breeders—and keep them in the lab at the same time, feed them the same food, sample the same tiny number of individuals as breeders, and repeat the process for the same number of generations.

If several generations later we find that in both of the populations selected for long horns, males end up with longer horns than before, and in both of the populations selected for short horns, males have horns that are shorter than before, then we can begin to rule out chance. In fact, I conducted my experiment on six distinct populations of beetles simultaneously. In two of the populations I artificially selected for longer horns in males; in two other populations I selected for shorter horns in males; and in the final two populations I chose the males at random. Keeping all these beetles supplied with food meant many, many mornings racing through the forest in search of monkeys. For six hundred sunrises I combed the damp understory, collecting bags of monkey dung to bring back to the lab so that I could feed the beetles of this huge experiment.[2]

• • •

Two years and seven beetle generations later, I had my answer. The weapons had evolved. Males in populations selected for longer horns now sported weapons that were proportionately larger than they had been before, and males in populations selected for shorter horns had weapons that were proportionately smaller. Each of these extremes differed from the populations that had acted as controls, and I'd shown convincingly that animal weapons can evolve fast.[3] But weapons weren't the only trait that changed; increases in horn size came at a price.

As weapons get big they also get expensive, and males with the longest horns now had stunted eyes. By the end of the experiment, males selected for longer horns had eyes that were 30 percent smaller than males selected for shorter horns. Stunted growth arises because of a limited availability of nutrients. Tissues require energy and materials to grow, and allocating resources to the production of one structure can mean that those same resources are no longer available for the growth of another.

Resource allocation trade-offs shape the development of all animals, but most of the time the effects are trivial. When animals begin to invest unusually heavily into particular structures, however, the effects of trade-offs become more pronounced. Weapons caught up in arms races get very big very fast, and in these species resources channeled into growth of the weapons can drastically impair bodily functions. In insects, this sometimes means reduced growth of other body parts.[4]

In dung beetles, we now realize, horn growth stunts the growth of eyes, wings, antennae, genitalia, and testes, depending on the species.[5] In exchange for fighting ability males suffer impaired visual acuity, flying agility, smell, and success in copulation—hefty prices for produc-

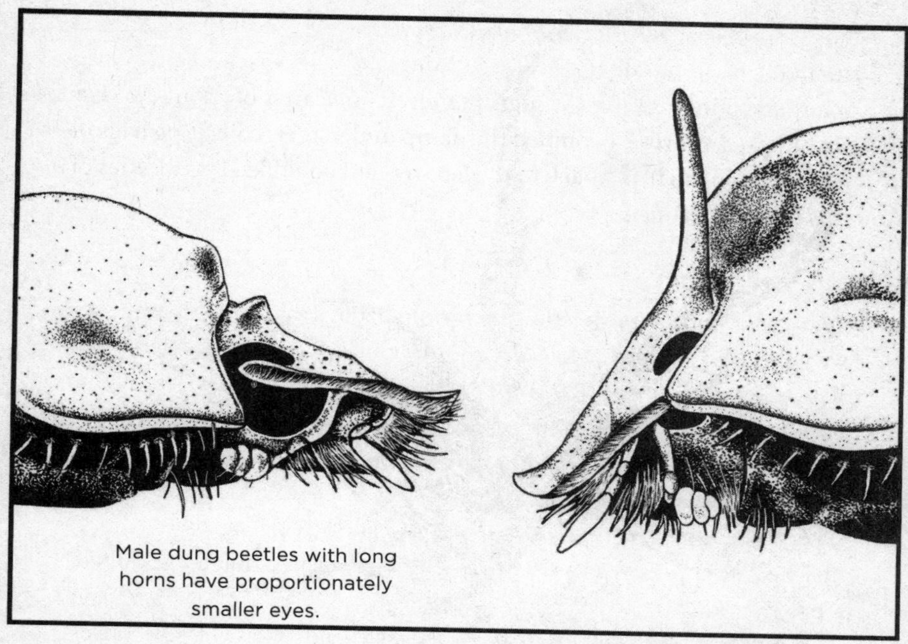
Male dung beetles with long horns have proportionately smaller eyes.

ing weapons—and similar trade-offs curtail a variety of species with extreme weapons. For example, giant rhinoceros beetles with massive horns have proportionately smaller wings,[6] as do stag beetles with the most distended mandibles.[7] Even male stalk-eyed flies face impaired testes growth when they invest in large weapons.[8]

Wing reductions in soldier castes of social insects are still more extreme. Big-headed soldiers in bees,[9] ants,[10] and termites have severely reduced wings and wing muscles, and, most of the time, these fighters are completely wingless. The price for winning fights is not merely impaired flight; it's an inability to fly.

• • •

Stunted growth of body parts is just one of many types of price that males pay for their weapons, and the bigger weapons become, the more they cost. Caribou antlers can exceed five feet in length and weigh more than twenty pounds—8 percent of the total body weight of a male. Moose antlers reach spans of six and a half feet weighing forty

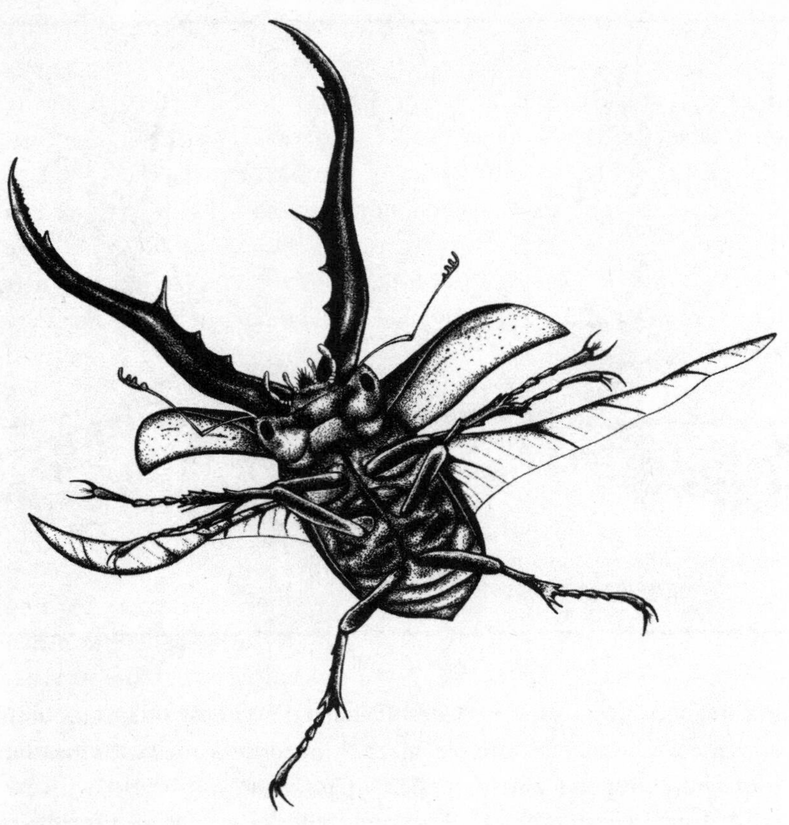

Stag beetles with big mandibles have smaller wings.

pounds, and the antlers of the largest extinct "Irish elk" spread more than fourteen feet across and weighed two hundred pounds. However, relative to body mass, the prize for largest animal weapons goes not to the elk, or even to a beetle, but to the fiddler crab, whose enlarged claw can weigh in at half the total weight of a male.[11] *Half* of the resources available to a growing male crab are allocated to weapon growth.

Not only are the claws bulky and heavy structures to build, they are energetically costly structures to maintain. Crab claws are not mere hollow threats; they are packed with powerful muscles that can crush the skeletal shells of rival males. Muscle tissue is incredibly energy demanding, because muscle cells are loaded with dense concentrations of mitochondria, the microscopic organelles responsible for converting stored nutrients and oxygen into usable energy. Mitochondria are often called the "cellular power plants," and inside muscle cells they provide energy needed to contract the muscle and close the claw.

Because of their many mitochondria, muscle cells are expensive to maintain even when they are resting, and males with big claws have the most muscle. Male fiddlers burn energy like crazy to keep their muscle cells alive. Resting metabolic rates of males with big claws are almost 20 percent higher than those of females (who lack enlarged claws) simply because of the costly muscles inside the claw.[12] Waving the claw around or using it to fight is even more expensive, and the bigger the claw, the steeper the energetic costs.[13]

Running with a bulky claw is also energetically demanding. Bengt Allen and Jeff Levinton devised a clever way to trick fiddler crabs into running on treadmills sealed inside airtight boxes. As the crabs ran, their muscles churned through contraction after contraction, burning oxygen and releasing carbon dioxide into the sealed container. Allen and Levinton measured the changes in concentration of these two gasses as each crab ran its treadmill race and, from this information, they calculated precisely the metabolic cost of running. Imagine running with a big bag of dog food in your arms, and it should come as no surprise that males with large claws burned more energy as they ran than males with smaller claws, or than females without enlarged claws. Actually, let's be fair to the crabs. To put this in proper perspective, imagine carrying a load that is equal to the weight of your body, as the largest of these claws can be (so, in my case, running with three fifty-pound bags of dog food and a cinder block in my arms). Good luck with that! Crabs with larger claws burned a lot more energy than crabs with smaller claws, and they tired much more quickly.[14] Male endurance on the treadmills suffered because of the extraordinary loads that they carried.

And the list of costs goes on. Female fiddler crabs have two frontal claws that they use to feed, plucking morsels of organic material from the sand and mud. This style of feeding is a delicate and tedious process, and the feeding claws of female fiddlers work incessantly as they scavenge. Males, on the other hand, have only one feeding claw because the other is grossly enlarged into the fighting weapon. The giant "major" claw of males is useless for feeding, so the males must make do with just the one. This can halve the rate of food intake for these already-energy-starved males, and males are forced to compensate either by spending more time feeding,[15] or by feeding faster,[16] with their remaining claw.

More time feeding means more time exposed to predators, and these exposed males are cumbersome, heavy, and awkward because of their claws—a dangerous combination. Several field studies of fiddler crabs have now shown that males suffer disproportionately at the hands of avian predators. In my favorite example, John Christy and his colleagues, including Patricia Backwell and Tsunenori Koga, studied a natural population of the fiddler crab *Uca beebei* on mudflats along the Pacific coast of Panama. They found that fiddlers were heavily preyed upon by great-tailed grackles, and that these birds had a quirky strategy for catching the crabs. Grackles often employed a "feinting reverse lunge" as they chased a crab. Instead of charging directly at a crab, they would charge to the side of the target crab, seemingly passing it by. As soon as they'd passed, they would whirl and lunge backward in a surprise diagonal feint that often caught the crab unaware. Grackles that did this were twice as effective at catching crabs as birds who simply ran straight at a crab and, remarkably, when birds employed the reverse-feint method, they caught almost exclusively males.[17] Male crabs, by virtue of their distorted major claws, were more conspicuous targets to the lunging bird. The result: in this population, male fiddlers suffered dramatically higher predation than females.

Increased exposure to the risk of predation turns out to be an almost universal cost to males for the burden of producing and wielding elaborate weapons. In fiddler crabs, higher predation can result from males being more conspicuous targets;[18] from their poor endurance and awkward, slower escape performance;[19] and even from predators actively seeking them out as preferred prey (the extra muscle inside the claw makes them more nutritious than female crabs).[20]

• • •

The best estimates of the costs of weapons come from studies of deer. Deer don't fit nicely into little plastic tubes, and they take a lot longer than dung beetles to complete their development. This makes artificial selection experiments problematic. But there are other ways to study sexual selection, and deer have proven to be ideal for many of these. For one thing, they are large, conspicuous, and relatively easy to watch. Deer are also easy to mark and follow as individuals, making it possible to track the fighting and mating success of dozens of different males, and the fawning success of comparable numbers of females. In addition, antlers are shed by males and regrown each successive year. Shed antlers can be weighed and measured and even ground up or incinerated to calculate the caloric and mineral content of the weapon.

Long-term monitoring of individual bulls can reveal how much time they spend foraging, chasing females, and fighting. Darting them with sedatives lets biologists have access to each male for an hour or so to measure their height, weight, and age (determined from the teeth), as well as count external parasites and sample blood to measure internal parasites and infections. Collecting this information before the breeding season—the rut—and then again after this season, and comparing these values, can shed light on just how expensive this whole mating process is to a male. In fact, rutting stags lose a stunning amount of body weight, and their physical condition plummets during the rut. Weapons, and the stamina, testosterone, and aggression that necessarily go with them, can be devastating to the health of a male.

Fallow deer (*Dama dama*) slide in neck and neck with caribou as the living species with the most extreme antler sizes. Fallow deer are native to Eurasia, and archaeological excavations in Israel suggest that they were an important source of meat for people as far back as the Paleolithic period (nineteen thousand to three thousand years ago). This species of deer was carried across central Europe by the Romans and introduced into the United Kingdom by at least the first century CE. Today, one of the best-studied populations resides in a rather unusual setting—an urban park inside the city limits of Dublin, Ireland.

Phoenix Park is no typical city park. It's one of the largest walled parks in Europe, with more than 1,750 acres of grasslands, hills, and forest. True, it is interspersed with tree-lined avenues and sidewalks, and the study animals do sometimes intermingle with an eclectic assortment of picnickers, joggers, and the occasional parade. But the deer in this population have lived their lives and died unmolested since the 1600s, and their rich and dramatic mating behaviors are on display for all to see.

The antlers of fallow deer flatten into giant, curved spoons with tines projecting in a ring from the outer edges like fingers splayed from the palm of a hand. A large buck may have as many as seventy tines fringing the perimeter of his antlers, and their full spread can reach wider than the male is long, a span of more than nine feet. For five weeks in September and October each year, rutting males wave these bulky antlers and bark from small display territories that they guard vigorously. They scream their throaty barks until they are hoarse, and they scrape at the soil, soaking each patch of exposed dirt with testosterone-laden urine to attract females and deter rival males.

Thomas Hayden and Alan McElligott have followed this population for more than fifteen years, during which it averaged between 300 and 700 animals. They were able to observe the rutting behavior and fighting and mating success over the entire lifetimes of 318 different males, recording who won the fights, who actually succeeded in mating, and how many offspring they sired. They also examined the costs that males paid: how much weight each male lost; how sick he ended up; and whether he was able to make up the lost weight before the onset of winter.

Not all males fared equally. In fact, in terms of reproductive success, an overwhelming majority of them failed miserably. Three-quarters of the males died before they were large enough—armed heavily enough—to succeed in guarding a territory, and 90 percent of the males never managed to mate with a female even once in their lifetime.[21] Of the males who did reach critical size and rank, most suffered catastrophic losses to their body condition as they battled for their tiny pieces of real estate, accumulating stress, slashes, parasites, and pathogens as they locked in battle after battle for territories that the females often ignored anyway.

Fighting for display territories and the females they might attract was a round-the-clock chore, with males averaging fights every two hours day and night for the duration. Males were not feeding during most of this time, and the displays and fights themselves were extremely energetically demanding. The result was that males typically lost more than a quarter of their body weight during this period. For a typical male, this was more than sixty pounds. By the end of the rut, most of the males were starved, exhausted, riddled with parasites, and nursing battle scars ranging from scrapes and bruises to broken bones and gashes. These battered bucks had only a few short weeks to regain their health and weight before the onset of winter. Males who failed to recoup these losses often died before the following spring.

Ron Moen and John Pastor used an entirely different approach to measure the price that male moose pay for weapons. By quantifying exactly how many milligrams of each mineral, carbohydrate, lipid, and protein an animal eats, and feeding this information into complex biochemical models of the physiology of vertebrate tissues, they were able to calculate precisely how much a male must shunt away from other body functions in order to sustain weapon growth.[22] What they showed was that antler growth in moose demanded 50 percent more energy per day from a male across the growing season, with peak demands reaching 100 percent (literally doubling the basal metabolic rate of all of the rest of the tissues of the body). Summed across the period of antler growth, this resulted in energetic demands as high as five times the energetic requirements of simply maintaining their bodies.[23]

Protein requirements of antlers were also high, but protein proved not to be limiting to these animals because the males could secure the extra protein they needed for antler growth through increased foraging. Interestingly, the crucial ingredients turned out to be calcium and phosphorus, both of which were absolutely critical to bone growth in the antlers, and neither of which was readily available to the animals as forage. In both moose and caribou, the calcium and phosphorus demands were so high that animals had to "borrow" these minerals from the other bones in their bodies in order to build their antlers. They could not get enough from their diet, so they pulled calcium and phosphorus from their own skeletons, reallocating them to the antlers. This is truly a form of deficit spending for these animals because it is not sustainable. The depleted skeletal reserves have to be replenished through feeding after the rut, and failure to do so is generally catastrophic.

All told, antlers in these animals turned out to be every bit as costly to a male as reproduction was to a female: the cost of building and using antlers was energetically and nutritionally equivalent to the cost of producing and nursing two fawns to weaning. Antler growth dramatically reduces bone mass overall, rendering males more fragile, more brittle, and much more prone to bone breaking. In essence, antler growth induces a seasonal form of osteoporosis exactly when animals are engaged in the most physically demanding and dangerous activities of their lifetimes. The rut is the worst possible time for males to suffer weakened and brittle bones, because this is when they put their strength to the test again and again in relentless brutal battles for dominance and reproduction. Antler-induced seasonal osteoporosis is no doubt part of the reason that, in many large deer species, fighting results in serious injury. Male moose suffer high incidences of fractured ribs and scapulas.[24] In red deer, a quarter of all reproductive males suffer bone breakage or other damage from battles during the rut, and 6 percent of stags are irreparably injured each year.[25] In bull moose, 4 percent of males are killed each year from injuries sustained during the rut and, across their reproductive lifetime, a third will die from injuries inflicted in battle.

In a clever extension of this research, Moen, Pastor, and Yosef Cohen applied their model to the extinct giant deer *Megaloceros giganteus*, otherwise known as the Irish elk. Technically, these deer were neither elk nor especially Irish. Close relatives of fallow deer, Irish elk were widely distributed throughout Europe, northern Asia, and northern Africa, until they eventually went extinct around 11,000 years ago. Most fossil specimens of this species are from Ireland (thus their moniker), from lake deposits of the Allerød period between 12,000–11,000 years before present. These magnificent deer produced the largest antlers known for any species, spanning twelve feet across in the largest bucks.

From fossil skeletons it's possible to determine the body sizes and proportions of these giant deer, and Moen, Pastor, and Cohen fit these

"Irish elk" had the largest antlers of any deer, shown here alongside a fallow deer buck.

values into their model to estimate how much the animals must have paid for the growth of their incredible weapons. Not surprisingly, Irish elk antlers appear to have been impressively costly, half again as much as the antlers of moose or caribou, and demanding almost two and a half times the basal metabolic energy requirements each day to grow. The calcium and phosphorus demands were severe, and seasonal osteoporosis was likely especially dire in this species. The time when Irish elk disappear coincides precisely with a period of rapid climate change called "the Younger Dryas." This would have lowered the quality of available food and made it even more difficult for males to replenish the calcium and phosphorus cost of their weapons.[26]

During the Allerød period, Irish elk lived in tall willow-spruce forests where high-quality forage was relatively abundant. However, pollen records show a radical shift in plant species composition at the end of this period, as temperatures plummeted during the mini ice age of the Younger Dryas. Relatively suddenly, the elk populations would have found themselves in tundra habitats with far worse forage conditions. It's possible the sudden decline in available diet exacerbated the material costs of male weapons by making it much more difficult, if not impossible, for males to replenish the calcium and phosphorus they borrowed from their skeletons each year to make the antlers. If true, then the extreme cost of male weapons may have contributed to the decline and eventual extinction of this species.

In the end, only the largest, fittest, best-armed males prevail in the competition for reproduction. For the fallow deer in Phoenix Park, one male in ten managed to mate at all, and the vast majority of copulations (73 percent) went to just 3 percent of the bucks. Such extremes in reproductive success—90 percent failure rates and extraordinary success by just a very few individuals—lead to intense sexual selection, and much of this is directed toward bulk, stamina, and big weapons. For the very best males, the reproductive benefits of investing in elaborate weaponry more than offset all of the costs combined. For all of the rest of the males, however, such armament extremes can be cost-prohibitive.

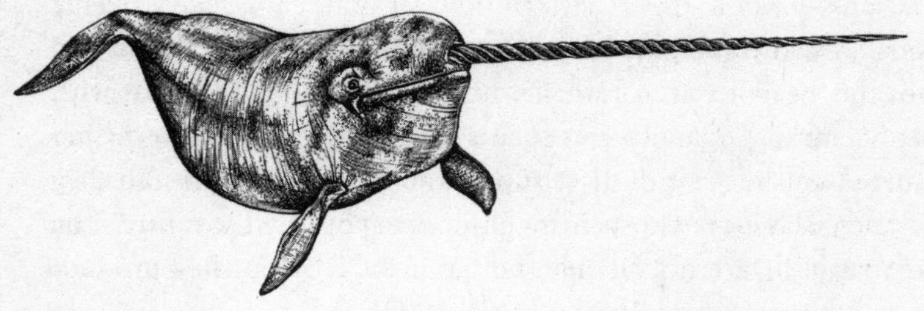

8. Reliable Signals

Appreciating when and why arms races erupt helps us explain why some species have big weapons and others don't—the "big-picture" patterns of animal diversity. But the science behind these races also provides insight into what happens *within* each of these species. Arms races unfold in the same basic way, proceeding through the same sequences of stages, in all species with extreme weapons. Parallel processes lead to shared properties, so much so that I could take information gleaned from weapons studied in one species, say a beetle, and use it to predict with frightful accuracy how weapons work in other species. Antlers in flies, forelegs in harlequin beetles, tusks in narwhals and elephants—these structures are alike in far more ways than simply being large. But to see this requires a shift in focus. Instead of considering variation in weapon size across species, we need to turn our focus inward, and look at variation between individuals in a single species.

Pick any one of these heavily armed species, and look closely at the weapons as they are expressed from male to male. Here, within populations, lurks another pattern: not all males produce extreme weapons. Measure a sample of one hundred males, and you'll find that most of the weapons aren't all that big. Sure, some males strut with monstrous racks. Organizations such as the Boone and Crockett Club keep meticulous records of these super-stud bulls and bucks. But they do this precisely because such magnificent specimens are rare. The majority of bulls are not Boone and Crockett caliber. They produce weapons, but they're middling.

Even though a history of sexual selection has led to evolution of extravagant weapons *in the species*, only a very few individuals actually achieve full weapon splendor, and lots of the males produce pathetic renditions of the structure. If males with the biggest weapons win in every sense of the word—they win the fights, they get the girls, and they sire the offspring—then why don't all of the males produce full-sized weapons? The answer is simple. They cannot afford them.

• • •

I could buy a forty-foot yacht if I wanted to. Okay, maybe I couldn't, but I like to think I could come close. An Azimut 40S has elegant lines and a stylish profile; two 480 horsepower engines; a spacious living room, master bedroom, guest room, and galley; and, of course, the latest navigation equipment and software. But it also costs $400,000. That's more than the value of my house and the fourteen acres of land it sits on. In principle, however, if I really wanted that boat, I could leverage my house against a loan, and for monthly payments twice my mortgage payments—basically, if I spent all of the income I had to live on and some of my retirement—I might be able to swing it. I wouldn't be able to feed my kids for the next decade, or take my dogs to the vet, or go to the movies, or do anything other than make payments on my shiny new boat. I wouldn't even be able to afford the gasoline I would need to drive it or a slip to park it in at the marina on Flathead Lake. But I'd have that boat, sparkling on a trailer in my driveway for all the neighbors to see.

Ted Turner, founder of CNN, lives just two counties over. I haven't met him yet, but I'd like to. I'm told he's the second largest landowner in the United States, and I know for a fact he's done incredible work restoring grasslands along the Rocky Mountain front, building a substantial herd of bison. Ted owns his whole county, and he could walk into a shop today and buy a forty-foot yacht in cash. He could easily buy two or three. But he's unusual. The mean income in this part of Montana is only $37,000 per year. For most of us, the cost of an Azimut 40S is prohibitive.

This may seem like Economics 101, but it illustrates a crucial point. Costs are not the same for everybody. Some of us pay a much steeper price for our toys than others. Technically, a 40S is a 40S. Ted Turner and I would each hand the dealer $400,000 for that yacht—exactly the same dollar amount. But the absolute value of the boat isn't the only thing that matters here.

Ted Turner and I aren't starting with the same amount of resources, and the cost of a 40S depletes a much bigger chunk from my resources than it does from his. Relative to what we each have available to us, that boat costs me a whole lot more than it does him. The bottom line: those of us with fewer resources pay a steeper price for luxury items.

When weapons evolve to extreme sizes, they get extremely expensive. They become big-ticket items, the yachts and Lamborghinis of the animal world. As a general rule, males produce the largest weapons that they can afford. But males differ in the relative sizes of the "pools" of resources they have at their disposal, and limited resources force the majority of males to produce substandard structures.

Of course, humans, too, have varying access to resources. A few among us are born to money, with family wealth passed from parent to child. Groomed in private schools with access to the best tutors and doctors, these kids grow up with contacts in the job market, fast-tracking them to lucrative careers in the best firms and companies. Others are born to hard times, living in low-income housing or run-down apartments. These kids may be forced to work early so that they miss a college education. As a result, they'll end up in low-paying jobs with little opportunity for advancement. Most of us fall somewhere in between, but as a population we differ widely in how much we have to spend.

A rich person can buy a house outright, without having to borrow. The rest of us must take out big loans, paying tens or even hundreds of thousands of dollars in interest to the bank, adding substantially to the cost of our houses. Banks charge higher interest rates for people with low income or poor credit scores, so the poorest among us actually pay the most interest, raising the cost of their houses still more. The sticker price for the 40S might not be different, but if Ted Turner pays cash and I have to borrow, I'll end up paying a much higher total for the yacht than he will. For example, if I borrowed at a rate of 5 percent interest, my total for a thirty-year loan would be $750,000, or $350,000 more than the list price of the boat! All of these factors exacerbate the differences among us, expanding the rift between haves and have-nots.

Animals are born with different resource pools, too. Elk calves born to the biggest, best-fed parents start life with an edge. They weigh more at birth, and have more stored nutrients and stronger immune systems. They have access to the best environments, including the safest, least-stressful places and the best food. Other calves begin life under duress, with poor-quality parents in poor physiological condition. They're born smaller and weaker, in substandard habitats. They grow more slowly and are quickly outpaced by the others. Because of their small stature they lose contests over food, weakening them still further, and the stress they experience increases the risk of infection. All of these experiences reinforce the differences in size that they started with, magnifying the gap between dominant and subordinate, large and small. Even subtle differences early in life compound as the animals grow, and by the time these calves reach adulthood they'll differ hugely in available resources. Only a very few will be able to produce the biggest, most extravagant weapons.

• • •

The real reason I couldn't afford that Azimut 40S is that not all of my assets are available for me to spend. Unless I'm grossly irresponsible to my family, I cannot go spending the equity in my house or my retirement account. Nor can I burn the cash I need to make my monthly mortgage and car payments, pay my taxes, or buy the food we need. Lots of my net worth is spoken for, and it's really only the income I receive above and beyond my mandatory expenses—my discretionary funds—that is up for grabs. In principle, I can spend this extra money however I choose. The problem is that once my fixed expenses are accounted for, there isn't very much left over, and my discretionary pool is far too paltry to spring for something as expensive as a forty-foot yacht. Discretionary pools differ from person to person far more radically than total resource pools do.

Animals work the same way. They have mandatory expenses that have to be covered first. Energetic demands for basal metabolic functions need to be met: things such as keeping the heart pumping, muscles contracting, digestive tract digesting, and brain thinking. All of these core functions burn calories and use nutrients, exacting a price from the animal, and these expenses are nonnegotiable. Default on them, and the animal dies. Only surpluses of resources—the biological equivalent of discretionary spending—are available for running and fighting, or building big weapons. This is why weapons begin growing so much later than other body parts. Horns, antlers, claws, and tusks all stay small during the course of development, and start getting really big only as males near adulthood—*after* they've already built their bodies—ensuring that mandatory expenses get paid first.[1] Only what's left over, if there is anything left over, gets shunted into weapons.

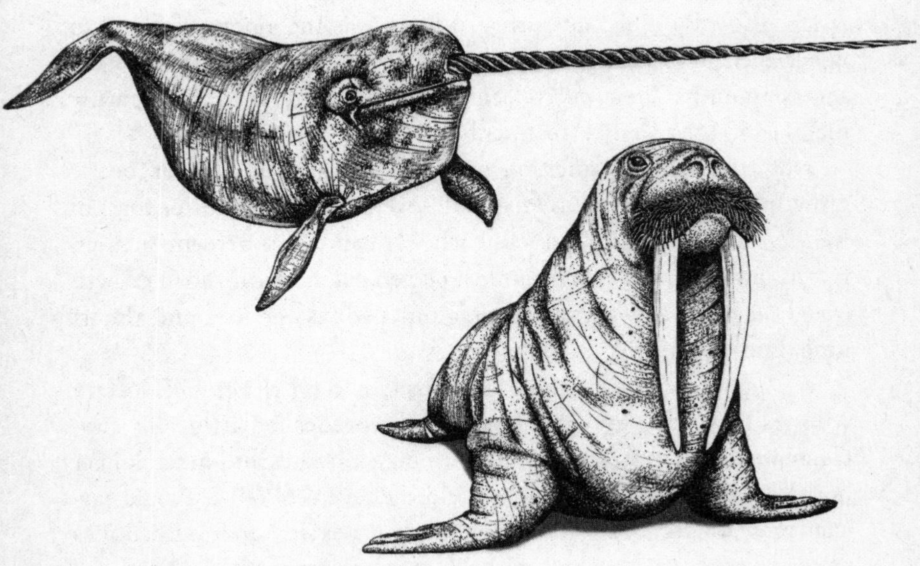

Weapons only get big after the rest of the body has grown.

Weapons are discretionary in another respect, too. They are not necessary for survival. Females, for example, fare just fine without them in many species, as do small males.[2] In stark contrast with the rest of the body, which must be built regardless, no weapon growth need occur at all. This means that the sizes of weapons should be far more sensitive to the availability of resources than other, mandatory body parts. My colleagues and I tested this a few years ago by perturbing the amount of food available to growing rhinoceros beetle larvae. By restricting access to nutrients, we experimentally altered the sizes of the resource pools available to developing males, allowing us to measure just how sensitive the different body parts were.

Rhinoceros beetles feed on decomposing logs. We made an artificial diet for them by fermenting sawdust in giant composters, mixing in healthy doses of year-old maple leaves. After about a month, the substrate turned chocolate brown and smelled like a wooded streamside on a rainy day, just the way the beetles like it. These are big beetles—a typical larva is the size of a mouse—and for this experiment we placed half of the larvae in pint-sized jars filled with substrate. We placed the rest of the larvae in gallon-sized jars fully filled with substrate. The only difference between them was the amount of food to which each larva had access. When adult beetles emerged from the jars several months later, we collected and measured them, comparing males from the two diet treatments.

Not surprisingly, nutrition had a pronounced impact on beetle growth. Genitalia in poorly fed males were 7 percent shorter than in well-fed males. Wings and legs each were about 20 percent smaller. Horns, however, differed by almost *60 percent*, meaning horn growth was three times as sensitive to nutrition as wings and legs, and almost nine times as sensitive as genitalia.[3]

All big weapons are exquisitely sensitive to nutrition. Like lottery winners upgrading to bigger houses, male beetles fed artificially supplemented diets grow into adults with bigger bodies and much longer horns. Remove the food, and you find the reverse. Well-fed male earwigs grow longer forceps than poorly fed males do,[4] and well-fed flies grow longer eyestalks.[5] The same holds true for deer antlers,[6] elk antlers,[7] and ibex horns.[8] Food is like income to an animal, filling its coffers so

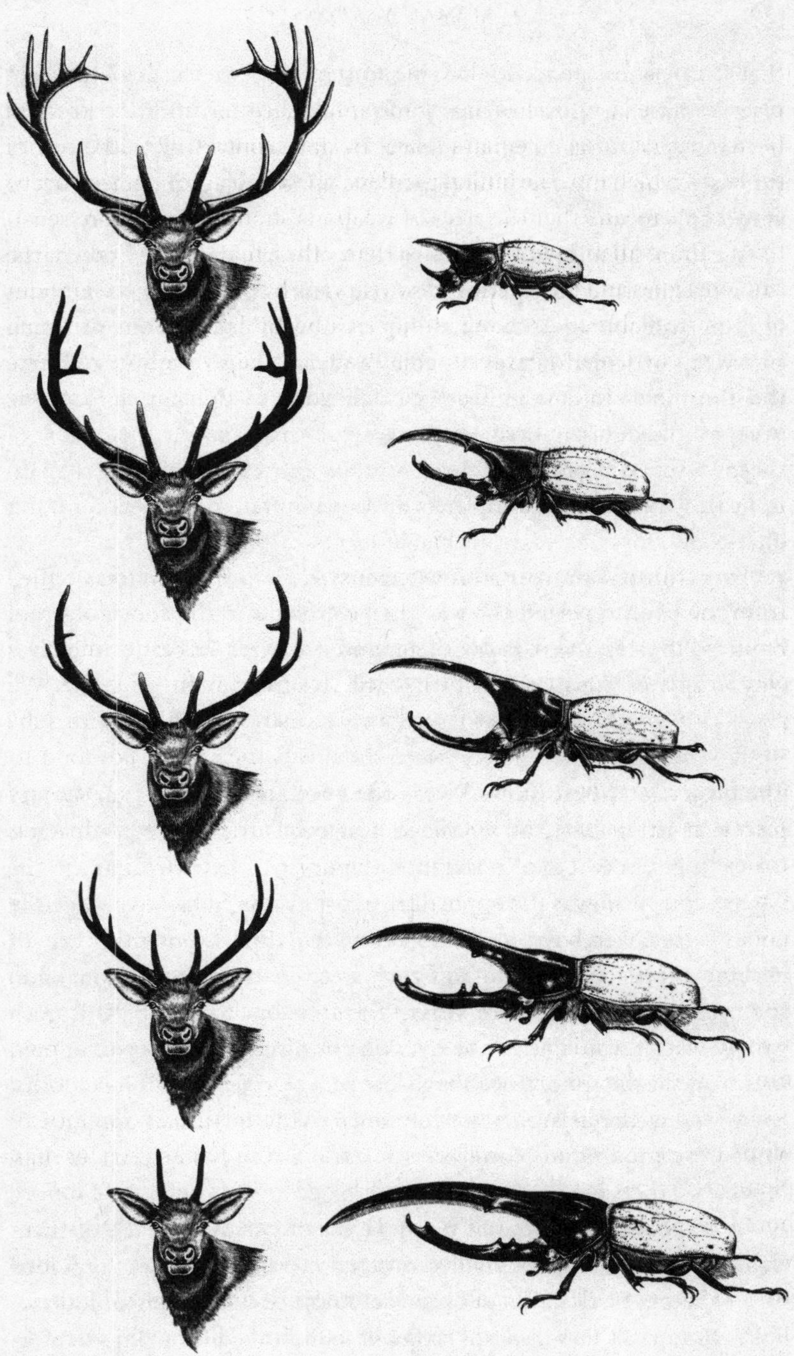

Same-age male elk and beetles differ in the relative sizes of their weapons, providing honest signals of fighting ability.

that it can later spend. Males able to sequester surpluses of nutrient reserves have large discretionary pools, and they can afford to produce big weapons. Other males have fewer resources to start with. Mandatory expenses may claim everything these males have, leaving nothing left to spend on weapons.

Weapon growth is extrasensitive to illness, too, for exactly the same reason. Infections drain resources from the surplus pool. Males fighting infections as they develop cannot afford to pour as much into weapon growth. Parasites gnaw away at tissues, pathogens battle the immune system, and all of this gobbles up stored reserves. Weapons and other discretionary structures absorb the brunt of these losses. Antlers grow much smaller in sick males than they do in healthy ones,[9] for example, as do Cape buffalo horns[10] and fiddler crab claws.[11]

Everything about weapons is expensive, from the resources pulled from the pool to permit their excessive growth, to the constant drain required to keep them, carry them, and use them in battle. Which is why weapon size is extrasensitive to the vagaries of life.

• • •

The biggest and best human weapons have always been exorbitantly pricey, unattainable to all but the richest few. During the Middle Ages, for example, the cost of a knight's armory was extraordinary.[12] The biggest cost of all was the opportunity cost. A knight had to be wealthy enough to never have to work. From the time a potential knight became a teenager, training to fight was a full-time occupation, often spanning a dozen or more years. The freedom to pursue this path simply wasn't available to the overwhelming majority of young men in Europe at the time, since they were indebted tenants of local lords. Even among the aristocracy who could swing it, not all apprenticeships were equal: some could afford to train with better masters than others.[13]

In combat a knight wore many layers of garments, each of them elaborate and expensive. Padded protection from shock was provided by an aketon, thick quilts of fibrous cloth filled with linen and horsehair. On top of this was a full jacket of mail, linked iron rings densely arrayed in fine overlapping rows designed to dull the slash of a blade. The best mail was tailored to the individual knight, so that it fit snugly around all joints without impeding movement. On top of this sat the armor proper, hinged plates hammered by metalsmiths into shapes that wrapped around shoulders, elbows, arms, and legs, as well as the chest and head.[14]

The quality of armor varied tremendously. The wealthiest knights had their own armorers who custom-shaped plates of the highest quality so that they fit perfectly. Others resorted to purchasing armor from local businesses. These suits were less expensive, but they were mass produced and "one size fits all." The poor fits of cheaper suits meant that they chafed when knights marched or rode into battle, and they restricted movement.[15] Finally, on top of the armor, knights wore elaborate and colorful tunics emblazoned with coats of arms or other identifying symbols, and the best of these were beautifully tailored and pricey.

Knights needed to purchase lances—lots of lances, since they shattered on impact during jousts—as well as pikes, swords, daggers, maces, and shields.[16] They also needed horses. A knight's warhorse was the most important, and by far the most expensive, tool of the lot. The best warhorses were bred for the task, tall, strong, fast, and reliable. The rarest and most highly prized were called "destriers."[17] These animals were trained from an early age to respond instantly to the slightest commands, and to walk in a perfectly straight line. They had to hold course despite extraordinary distractions: screams and howls of battle and foot soldiers rushing them swinging axes and clubs. A horse that balked at the moment of battle was a dangerous liability. Breeding and training the best horses cost a fortune, and the wealthiest knights had three or four in their retinue.

Warhorses had to be equipped with armor, which cost even more than the knight's armor did, since there was so much more of it. Padded underlayers, mail, plate armor, and extravagant outer tunics were the norm. Here, again, the best armor was custom-built for each animal, so that it didn't chafe or impede movement.

Knights spent months of the year away on campaigns, and they brought with them fancy tents, trunks with clothing and gear, carpets, kitchens, cookware, and furniture. Strings of packhorses carted the freight, and additional horses carried squires, apprentices, servants, and cooks.[18] Everything from the breed of the horses to the styles and sizes of tents, clothing, armor, and entourages separated the best from everyone else. Like the horns of beetles and the antlers of elk, the quality and extravagance of a knight's shining armor revealed his status and wealth, and, because training, ease of movement, and protection all matched the price of his armor, a knight's appearance also revealed his fighting strength.

Training, horses, weapons, and armor were extraordinarily expensive, unattainable for all but the wealthiest sons of noble families.

• • •

Because differences in animal weapon sizes reflect the health, nutritional history, overall condition, and genetic quality of each male, they're a meaningful signal—a visual proxy—for fighting ability. Of course, other body parts differ, too. The best bull elk stand taller, and have bigger heads and longer tails, than poor-quality males. But weapons make better signals for two reasons.

First, they're vastly more variable than other elements of an animal's arsenal. No elk stand zero inches tall, for example. All elk have bodies and muscles, but there are plenty without antlers. Weapon size spans all the way from zero to enormous, resulting in a bigger range from male to male than in other body parts.[19] It's far easier to discern six-foot-long antlers from a pair of six-inch spikes, than it is to tell that one male stands a few inches taller than the other. Even subtle differences in the body sizes or fighting abilities of males become amplified into pronounced differences in the relative sizes of their weapons.[20]

Second, these structures are huge. Weapons are massive and conspicuous outgrowths, biological billboards advertising the quality of a male for the world to see. Best of all, these billboards are honest. A puny bull elk could no more fake a giant rack of antlers than I could buy an Azimut 40S.

Suppose for the sake of argument that I did go buy that boat. Say I threw caution to the wind, and I bought it. I couldn't afford to register it, or buy fuel to drive it, or repair it if it got dinged. Similarly, a poor-quality bull that somehow grew big antlers wouldn't be able to use them. He'd lack the size, strength, energy reserves, and stamina necessary for wielding his antlers effectively in battle. The effort would be futile.[21]

The price of big boats, like big weapons, increases exponentially with size, and people tend to buy the biggest boats they can. Next time you stroll through a marina or harbor, pause for a minute. Take a gander at the variation in the sizes of the boats people amass for their recreation. Little itty-bitty boats, in-between boats, big boats, and, occasionally, parked by itself way out in the harbor, or on its own dedicated pier, a really, really big boat—a 150-foot floating mansion of a yacht; the top of the pack. Yachts are status symbols for a reason. Like weapons, their size clearly and accurately reflects the resource pools available to their owners.

Males invest as much as they can in their weapons, but not all males have the same amount available to spend, and weapon sizes differ wildly as a result. Ultimately, though, this comes in handy. Because weapons display crucial information about the health, status, fighting ability, and overall quality of a male, and because they're so visible, they make it easy for rivals to assess each other before allowing confrontations to escalate into dangerous battle.

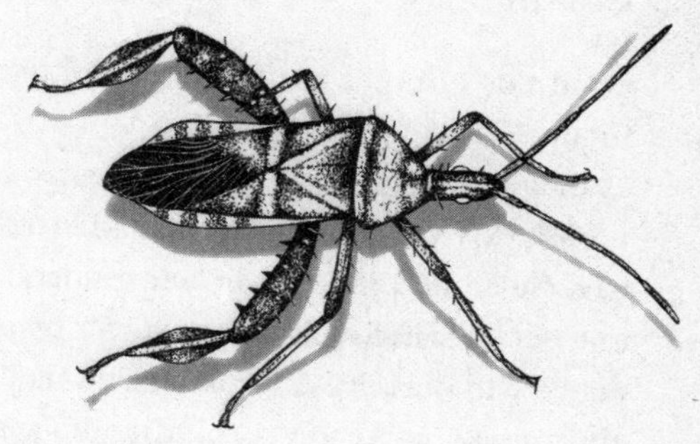

9. Deterrence

Suddenly I was awake, the moist floor of the tent ballooning around my face. As I sat up on my knees the nylon fabric whooshed to fill the void where I'd been, squeezing up gently around my wrists and knees, gurgling like a giant waterbed. I could hear rain smacking onto the roof and sides of the tent. A storm had blown in and it was pouring; but that wasn't it. There was something else. Waves. We were surrounded by waves. Somehow, the tide had risen farther than expected and waves crashed into our tent, engulfing us as they raced up the shore.

We hopped into the darkness, rain streaking down in sheets, to find ourselves shin deep in warm water. The three of us grabbed corners of the tent and yanked, uprooting the whole thing in a jerk, frame bending precariously from the weight of our sleeping bags sagging in the bottom. Holding the dripping contraption high enough to clear the surf and laughing hysterically at the ridiculousness of our situation, we raced for higher ground. (We learned the next week that a hurricane hundreds of miles to the north had triggered a record-high tide that night.)

Soggy exodus aside, this had been one of the most magical nights of my life. My wife, Kerry, our friend Lisa, and I had read about this pristine Costa Rican beach in our guidebook. The five-mile hike kept all but the occasional hardcore surfer away, so we had the place almost entirely to ourselves. The scenery was postcard perfect, tropical forest abutting white sand and blazing blue water, palm fronds fluttering in the breeze. After exploring in the afternoon, we'd set up camp at the edge of the trees so our tent peeked out over the water. But the real treat came with darkness.

With no lights or buildings for miles in any direction, the night was completely black and stars in the sky dazzled. The surf dazzled, too. Churning waves glowed blue-green from phosphorescence in the water. Each wave lit up as it curled, bright sheets of brilliant green lifting to the sky, before collapsing into the shore and sending glittering foam across the sand. All the way up the beach, waves lit the night and the sand shone green. As we skipped through the darkness our footprints and the soles of our feet glowed, so we choreographed neon paintings in the sand with our tracks.[1]

The beach came alive in another way, too. Ghost-white crabs skittered everywhere, nickel-sized bullets shooting from underfoot as we danced. At the water's edge the little crabs fed on detritus in the glowing froth, bits of phosphorescence sticking to their bodies. Inshore from the water, thousands of thumb-sized holes sprinkled the beachscape. Spaced maybe a foot or so apart from one another in a dotted grid, they covered the beach. Beside each burrow hovered a crab, poised and ready to dart for cover when predators ventured too close. Thousands of additional crabs flurried about, approaching and occasionally challenging owners of the burrows. If we stood absolutely still, they darted between our legs and over our toes.

We were witnessing a spectacle few ever notice even though it unfolds on tropical beaches across the Pacific, Atlantic, and Indian Oceans. Night after night, day after day, tiny crustacean fighters face off in the sand. Ten thousand battles unfold each night on this beach alone, and beaches just like it all over the world. Ghost and fiddler crabs aren't hard to find, or to observe. Anyone spending any time on a beach has seen them, but few of us take the time to watch what they do.

One person who does watch is John Christy. For more than thirty-five years, John has studied the behavior of fiddler crabs, spending countless hours watching them fight and court on strips of beach and tidal flats of Panama. As a doctoral student in the 1970s, John lived by himself on a tiny island in Charlotte Harbor, Florida, with no one to talk to except his pet smooth-billed ani and the crabs. Dirt-poor, as graduate students almost always are, he convinced the manager of a nearby field station to let him live on the island for free. The only building, a premade one-room aluminum box lacking electricity and plumbing, became his home. Two thousand feet long and a thousand feet wide, the smudge of land called Devilfish Key was hot, humid, and thick with mosquitoes—a far cry from our idyllic Costa Rican paradise. But its tiny beaches were covered with millions and millions of crabs.

Nobody ever visited Devilfish Key, so John could set up experiments without interference. He placed hundreds of colored flags in the sand, marking and mapping the locations of burrows. He excavated tunnels to see how deep they went and discovered females tending broods inside some of them. Super Glue had just been invented, so he stuck tiny colored tags onto the backs of five hundred crabs, recording the body size and claw size of each animal and watching to see what they did. He recorded who approached whom; how they interacted when they faced off at close range; whether they fought; who won if they fought; and who succeeded in breeding.[2] Most of all, John wanted to see what they did with their extraordinary weapons.

What he found was they wave them. Up and down, up and down, again and again they raise their claws high and drop them. Dozens of times each minute, thousands of times per hour, hour after hour. When contests get intense, claws clearly function as weapons—dangerously strong claspers capable of inflicting real harm. Yet they also appeared to serve as signals. In fact, for every few minutes of outright fighting, males spent dozens of hours waving. Most of the time,

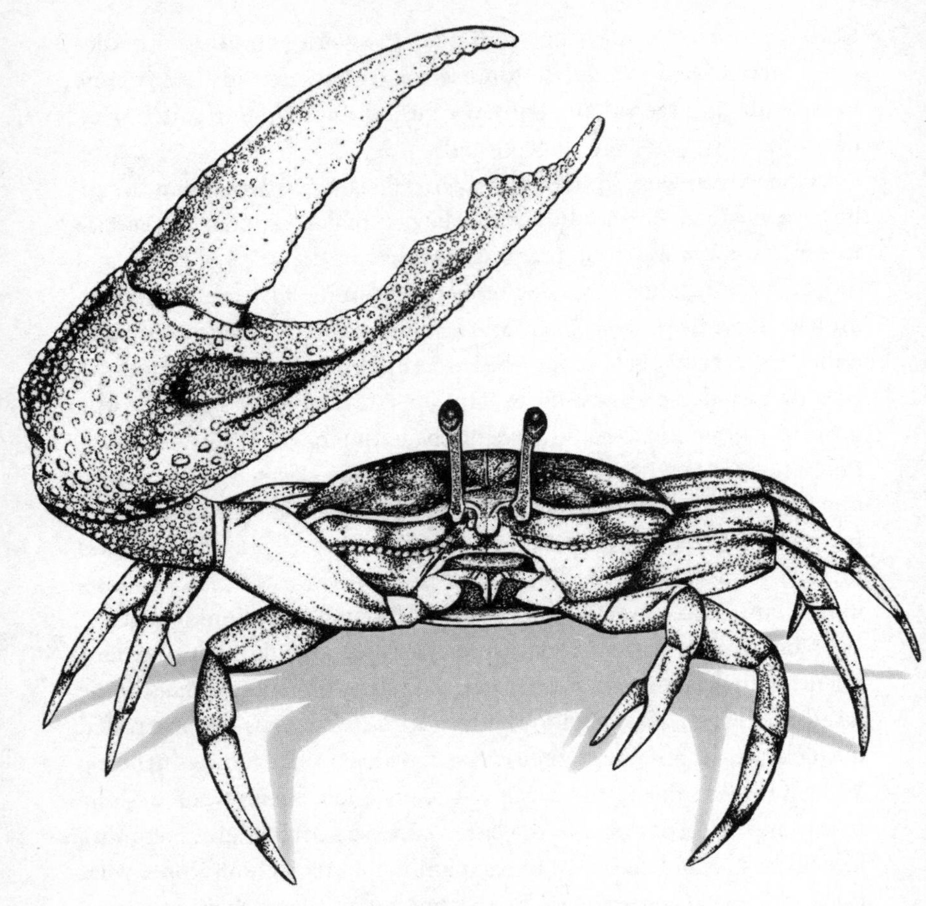

Male fiddler crabs wave their claws, using them as deterrents first and, only occasionally, as actual weapons.

fiddlers were employing their claws as warnings rather than instruments of battle. They were using them for deterrence.

• • •

The best way to tell if you can beat opponents in battle is always to fight them. Charge with everything you've got, hold nothing back, and, at the end of the brawl, you'll have your answer. The problem with fighting is it's dangerous. Sometimes the risk is simply a by-product of being distracted. Crabs are well protected from one another, since their exoskeletons are like armor, but fighting crabs are distracted crabs and they make easy targets for gulls and grackles. Other times distraction is deadly because of where the fight takes place. Bighorn sheep and ibex smack heads on narrow ledges of steep cliffs, and a single misstep can be disastrous. Even a broken leg is fatal, so males pay constant attention to their opponents and their footing as they fight.

Most of the time, the fight itself is dangerous. Elephant seal bulls are always covered in gashes. I've seen bulls with six-inch slabs of fat and skin dangling from cuts inflicted in battle. Tines on horns and antlers catch and parry an opponent's thrust, but tines get tangled, and bulls whose weapons lock die helplessly intertwined. Rhinoceros beetles with holes in their cuticles are commonplace, punctured by barbs on a rival male's horns. Jabs and stabs from tusks can shatter bones, inflict deep stab wounds, and trigger infections. Almost every fight brings with it some risk of real harm, and males with battle scars are so ubiquitous in the animal world that we tend to take them for granted.

What if there was a way to tell who was going to win before entering into dangerous battle? If a male could discern that he was likely to lose beforehand, he might opt to just walk away. A male ceding a fight outright forfeits the chance of mating, but he lives to fight another day. Walking away when he might have won would be crazy, but ceding fights he would have lost saves time, energy, and risk. This is particularly true for the smaller contestants in poorer condition, since they're the ones most likely to be maimed in the fray. The trick, of course, is to accurately predict who's likely to win, and to do that, rivals must have an easy way to size each other up. They need a conspicuous indicator of the fighting ability of potential competitors.[3]

When fiddler crabs size up rival males, they look at their claws. For one thing, they're huge and, in many species, brightly colored, making them supereasy to see. They're also the most variable part of a crab, ranging from tiny to extreme and including everything in between. Finally, like all big weapons, claw growth is sensitive to parasites, disease, and nutrition, so the male with the bigger claw really is the most likely to win.[4] This makes claws worth paying attention to, and males use them to evaluate one another before battle.

• • •

Fiddler fights revolve around burrows. Resident males defend ownership of a burrow, cleaning and widening and perfecting it each time they have a moment to spare. Females visit a succession of males, inspecting each burrow before picking just the right one. After choosing, they mate with the male and lock themselves belowground for several weeks while their broods develop. Females are consistent in their choice of burrows, picking tubes of the right width and depth, positioned just the right distance from the water—high enough to escape flooding by the highest high tides, but low enough that the bottom of the burrow stays moist.[5]

Males battle for burrow ownership, guaranteeing that the biggest, best-armed males end up residing in the most attractive burrows. At any point in time, the beach will be blanketed with guarding males, stationed as if tethered by threads to their tunnels, waving claws like banners in the air. But beaches are full of wandering males, too, and animals cycle between these two states. Guarding males cannot eat, since their food lies at the waterline farther down on the beach. They live off of stored nutrient reserves, depleting them a little more each day. Eventually, even the best males run out of steam and are forced to abandon their burrows to go feed and refuel. The instant they leave others will claim their burrows, so they'll have to work their way back up the beach as wanderers, challenging males in attempts to claim new burrows as they go.

The sheer numbers of crabs—hundreds of thousands per beach—and the constant turnover between wanderers and guards, result in an astounding number of face-offs. Each resident male faces hundreds of challenges per day, and evictions and turnovers are commonplace. These many encounters would be dangerous indeed if they all involved pitched battles. But they don't. Most confrontations settle long before they escalate that far.

Imagine the beachscape from the vantage of a wandering crab, eyes perched on stalks a mere inch above the sand. Everywhere he looks are claws jerking up and down, bursting above the horizon and back again. Over and over they rise and fall, surrounding him with an incessant barrage of motion. The wandering male doesn't simply march up to the first burrow he encounters. Nor does he approach at random. As he weaves between waving males, he sizes them up—bigger claws reach higher than smaller claws—advancing only toward burrows guarded by males whose claws are equal in size or just a bit smaller than his own.[6]

This is remarkable not only because it implies males can tell how big their own claws are (how else could a male discern at a distance that some claws were too big?), it also means that males opt out of most confrontations from afar. An overwhelming majority of contests end before they ever begin, without anything even resembling a fight. A mere glance at a big claw is sufficient to deter smaller males.

Only when a male finds an appropriately sized rival will things advance to the next level. As a wandering male approaches, the resident turns to meet him head-on. The guarding male extends his claw forward, and the wanderer pushes back against it in a gentle sequence of shoves, claw to claw.[7] If the intruder guessed poorly on his approach and the resident's claw is too big, he'll back down at this stage. If not, then the pushes get a little rougher. Each male rubs his claw along the length of his opponent's, sliding claw against claw in the process. Here, too, many males opt to depart peaceably. Unless they're very evenly matched, the smaller male walks away.[8]

If both males hold fast past this stage, then the confrontation escalates once again. This time it includes much more forceful shoves and clasping—squeezing with the full strength of the claw. Finally, if neither backs down, males enter into unrestricted combat, attacking with powerful strikes and grabs in a fierce war of attrition. The resident may back into the safety of the burrow and shield himself while the intruder hammers his claw against him and continues to strike until one of them finally gives up and flees.[9]

Fights reaching the pinnacle of this sequence are furious and all-consuming, energetically demanding and dangerous. But fights this intense are rare. Given the incredible number of male-male encounters that take place on beaches each day, it's remarkable how few reach actual battle. For every contest settled in a full-blown brawl, hundreds are resolved peaceably. Fiddler crabs have the largest weapons relative to their size of any living animal but, because claws act as deterrents, they almost never have to use them in fights.

• • •

Whenever there is an adequate cue that conveys information about a male's fighting ability, it will always benefit other males to pay attention.[10] There really is such a thing as choosing battles wisely. Males who fight every battle to the fullest, attacking indiscriminately and never holding back, end up mangled and exhausted, or dead. By assessing the abilities of rivals before fights get costly, and attacking only when the likely payoffs exceed the risk of injury or failure, males allocate resources more efficiently.

Like fiddler crabs, male bamboo bugs use their weapons as deterrents first, and only occasionally as instruments of battle. Huge hind legs bulge out from their flanks. Thick, strong, and armed with sharp spines, these legs can squash a rival male, crushing and puncturing cuticle in the process. Males guard harems of females crowded onto new shoots of bamboo, shepherding them together and facing off with intruder males. As rivals approach, guarding males wave their big legs in the air, flashing them in the faces of the intruders. Usually this is a sufficient deterrent. Only when males are very evenly matched do confrontations progress into actual battles.[11]

Ibex have the longest horns of any ungulate, yet they, too, almost never fight. Males size one another up constantly, strutting and some-

Bamboo bugs wave their weapons to deter rivals, and only escalate to fights if they are evenly matched.

Ibex rams size each other up, comparing weapon sizes; most confrontations end without escalating to battle.

times running side by side, and they spar. But smacking heads in full-blown battle is extremely rare.[12] Caribou are similarly cautious. One study followed more than 11,600 male-male contests over two years; only six escalated into outright battle—less than one-twentieth of 1 percent.[13]

In one of nature's more amusing paradoxes, the most extreme weapons are also the least likely to be deployed in pitched battle.[14] The biggest, best-conditioned males wield massive weapons capable of deterring most rivals on the spot; the mere presence of these weapons is sufficient to dissuade all but the biggest competitors. For the rest of the males with intermediate or small weapon sizes, battles are engaged only when rivals are equitably matched.

• • •

Deterrence is an integral and intuitive stage in the unfolding of an arms race. As weapons increase in size they get more expensive.[15] Fewer and fewer males can afford to pay the price, widening the gap between rich and poor. By pulling the extremes farther apart, evolutionary increases in weapon size magnify the rifts between haves and have-nots. Zero stays zero, but the biggest weapons keep getting bigger. As signals, weapons become more honest and more visible simultaneously, fueling the evolution of deterrence.

Deterrence, in turn, feeds back on the arms race, driving weapon evolution forward at a faster and faster rate. The moment weapons start functioning as cues of fighting strength, a whole new incentive for extreme size arises. Now, males with the biggest weapons win for two reasons: because they defeat rivals in battle, and because their weapons are big enough to end most confrontations without a fight.[16] Fight costs saved by deterrence compound already-impressive rewards for males with the largest weapons.

Of course, males aren't the only ones who pay attention to weapon size. In many species females study them, too.[17] And why not? They're obvious and reliable advertisements for the quality of a male. Female crabs are more likely to approach males with bigger claws,[18] and they prefer claws with bright colors.[19] Female stalk-eyed flies prefer males with longer eyestalks;[20] female earwigs prefer males with longer forceps;[21] and red deer[22] and topi[23] females prefer males with bigger antlers and horns. (Attracting the girl is yet another reward for having impressive weapons.)

Finally, the nature of deterrence—the way males assess each other before battle—reinforces the one-on-one nature of male contests, strengthening the conditions conducive to an arms race. Repeatable sequences of assessment behavior work like a branch or a tunnel in the sense that they force males to duel. Even if fights occur in the open, as they often do with caribou and antelope, the stages of sequential assessment align contestants so that full-blown battles always end up head-to-head and one-on-one, ensuring that when outright battles do occur, the better-armed male wins.

As weapons get bigger they select for increasingly elaborate deterrence, and deterrence, in turn, selects for bigger and bigger weapons. Arms races and deterrence push each other forward, escalating in an evolutionary spiral. Like a figure skater pulling her arms to her chest, the evolution of extreme weapons gets faster and faster.

• • •

In its prime, the British Empire controlled a fifth of the world's population, with colonies and territories on every continent. This was a remarkable accomplishment for a tiny island nation, and it would not have been possible without the absolute supremacy of her navy.[24] Throughout the eighteenth and nineteenth centuries, the Royal Navy was the largest in the world, filled with fleets of sailing fortresses—magnificent sailing ships with tall masts and solid oak hulls stuffed with cannons.

The cost of building warships was enormous. The hull of a single seventy-four-gun ship required 3,500 oak trees—mature hardwood trees at least a hundred years old each—and a one-hundred-gun ship needed almost six thousand trees.[25] With European nations largely deforested already, these hardwood costs were incredible; only countries with extensive networks of shipping routes and colonies could afford to import the necessary lumber, and, most of the time, they simply assembled the ships in their colonies. Add in the costs of shipyards, engineers and shipwrights, labor crews, cannons, rigging, and trained officers and crews, and the price of the biggest warships was way out of reach for hundreds of the world's nations. Fleet size and ship size became signals of a country's fighting ability—perfect cues for deterrence.

Just like crabs, when navies clashed, ships sought out rivals of similar size. Fleets lined up single file, leading with their largest ships and assembling them in descending order down the line.[26] Opponent fleets did likewise, matching the largest ships against one another. Navies with the greater numbers of bigger ships had an edge since their string of "heavies" extended farther down the line. Even when battle lines devolved into frantic melees, like still sought out like. Big ships could always beat smaller ships, but their bulk and weight slowed them down, and smaller ships shied away. Medium ships escaped larger ones, but were too slow to catch the small ones, and so on. Warships inevitably chased down rivals of comparable bulk and speed.[27]

The flagships of these navies were the "first-rate" ships of the line—the most massive ships in existence in their day. First-rates were potent weapons, capable of shattering the hulls of lesser ships with single broadsides of cannon fire. They were also stunning to look at—paragons of power. One glance at these splendid ships was enough to know just how effective they could be in a battle. Simply sailing a ship this big into troubled harbors could quell uprisings or settle disputes anywhere in the world.[28]

At a time when most of the world had no first-rate ships at all, Britain kept dozens and dozens in her fleet. During the Napoleonic Wars, for example, Britain had 180 warships of seventy-four guns or larger.[29] The Spanish, Dutch, and French navies all vied for power with Britain at one time or another,[30] but none could match the British navy and, by the end of the Napoleonic Wars, Britain alone ruled the seas. For most of the nineteenth century, a period called the "Pax Britannica," the majority of local conflicts were settled without escalation simply because of the presence of a British warship. As with crabs and caribou, big weapons and deterrence ushered in a window of relative peace.

The cost of state-of-the-art weapons is still staggering today, and only the richest can afford the best. More than one thousand feet long and displacing one hundred thousand tons of water, *Nimitz*-class nuclear-powered aircraft carriers are home to ninety fighter jets, batteries of antiaircraft missiles, and more than six thousand crew members.[31] A ship this size costs $4.5 billion to build, not counting the aircraft. Modern fighters such as the F/A-18 E/F Super Hornet cost $67 million apiece,[32] bringing the total ship cost closer to $10.5 billion. Add the price of training and paying six thousand military personnel, and the cost climbs still higher. On top of this, each carrier by itself is vulnerable to attack, so they never travel alone. Carrier strike groups consist of a carrier plus a miniarmada of support craft, generally two guided-missile cruisers, between two and four antisubmarine and antiaircraft destroyers, and, often, a submarine. The purchase cost for a carrier strike group exceeds $20 billion, and one recent study estimated operating costs of maintaining a strike group at $6.5 million per day.[33]

The United States has ten *Nimitz*-class carrier strike groups; no other nation has anything even close to this. We employ our navy in ways similar to nineteenth-century Britain. Massive, powerful, and prohibitively expensive, our carriers function both as weapons and deterrents, portable projections of military power shuttled like chess pieces to stabilize troubled regions.

10. Sneaks and Cheats

During my final year in Panama, after dawn dashes into the forest to find monkeys and in between generations of the artificial selection experiment, I spent my days in the dark under a tent of thick cloth strung from the ceiling of my office. This time, my objective was simply to watch. Anything and everything the beetles did, I wanted to see—no one had ever studied their behavior before, and I was eager to see how the males used their horns.

The problem was, everything interesting happened belowground. I didn't have thousands of crabs fighting at my feet or jacanas prancing on mats of floating lettuce. These guys disappeared into pencil-sized burrows in the clay. At the end of the nineteenth century, the French naturalist Jean-Henri Fabre overcame a similar challenge while studying the underground breeding behavior of a European dung beetle species. He secured a pie tray with a hole in it on top of a vertical soil-filled glass tube.

A hundred years later I'd graduated from glass tubes to glass sandwiches. I built a series of "ant farms," packing soil between panes of glass. Instead of pie trays, I used Plexiglas boxes with slits cut into the floors, securing one to the top of each ant farm. When the beetles tunneled, they had no choice but to excavate in between the panes of glass, permitting me to peek inside. Bright lights bothered the beetles—the insides of tunnels don't normally get much sunlight—so I had to simulate darkness. Fortunately, dung beetles are like harlequin beetles in that they cannot see the color red. Using red-filtered lights, I could watch without disturbing them from inside my black cloth tent.

For four-hour stints I scribbled notes, squinting in the dim light while watching pea-sized beetles tussle inside tiny tunnels. Heat from the lamps cooked the little tent, and the smell of must and manure was overpowering. But the beetles thrived inside the glass sandwiches. They fought and mated and provisioned their young, and I got to see it all.

It was immediately clear that males use their horns in fights. This was not surprising, but it was exciting to see nevertheless. Fights were wonderfully chaotic. The guarding male braced himself, leg spines lodging into the soil walls of the tunnel, the intruder pushing into him, forcing himself downward and twisting with his head and horns. The horns of the males would enmesh as beetles pried with their heads and pushed. If the males were evenly matched and the fights escalated, they would become increasingly frenetic, with males forcing the tunnel wider and flipping past each other. Back and forth they'd switch positions as first one and then the other managed to slip into the prized inside spot. Sometimes the dueling beetles backed all the way down to the female, slamming against her in the fight. Other times they'd tumble outside the tunnel onto the surface. During the craziest fights I'd lose track of who was who but, at the end of the scuffle, it was almost always the male with the smaller horns that ended up leaving.[1]

After I'd watched enough of them, the fights became a blur, playing out the same way every time. Watching the winners wasn't very exciting. In the end, it was the losers who surprised me. If a big male lost a fight he'd storm off in search of another tunnel and another challenge. In the field he'd have to travel only a half inch or so before reaching the next tunnel; in my cages he had no such luck, so he'd walk in endless circles around the perimeter of the Plexiglas boxes. But tiny males did something very different. After getting booted they'd only go a short distance away—maybe a half inch—and there they'd begin to dig their own tunnels. Excavating tunnels is typically a female behavior, but here these little males were making new tunnels, right beside the guarded tunnels.

I got very excited when I first saw this, thinking the little male might sneak back into the main tunnel belowground. But all he did was sit there. For hours he just sat, and I sat, until I started to get restless. Of course, it was the moment when I left to use the bathroom that he did it. I came back to find the whole thing over. The little guy was back in his tunnel, but I could tell from the chamber that he'd excavated a side tunnel, drilling right across into the main burrow. I started setting up five or six chambers at a time, all with mixtures of large and small males, and sure enough, I finally saw the sneaking firsthand. After sitting still for hours, the little male suddenly stirred, pushing into the side of the main tunnel and shooting down the shaft to the female. He

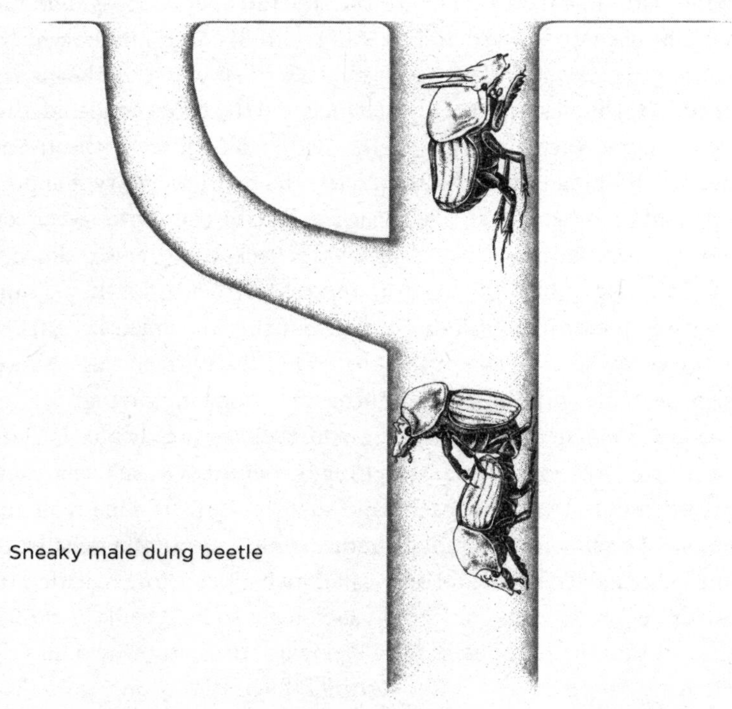

Sneaky male dung beetle

could mate with the female and bolt right back out again in just a couple of minutes, while the guarding male blocked the entrance above, oblivious to the intruder.

When I told my doctoral committee about the side tunnels they challenged me, pointing out that these beetles were digging in skinny glass sandwiches, a two-dimensional universe. Where else were the little guys going to go? Of course they'd bump into the main tunnel in an ant farm. The real question was whether they did the same thing in the field. Armed with tubes of warm silicone caulking, I squirted white goo into tunnels in the forest. Pieces of monkey dung aren't very big—maybe the size of a silver dollar—and it was not uncommon for there to be entrances to ten or twenty separate tunnels crammed into the soil beneath them. So I filled them all with silicone, then dug the whole thing up and carted it back to the lab where I could gently wash away the soil, revealing casts of the tunnels.

Not only did side tunnels happen in the wild, they happened often. It was clear from the casts that sneaky males could hit four or five tunnels with each of their horizontal shafts.[2] Now, I understood why big males periodically patrolled their tunnels, and I was beginning to understand why small males don't produce horns. In this species, like many of the tunneling dung beetles, the largest males all produce a pair of long horns, but the smaller males do not. They don't even have intermediate horns. Instead, they appear to shut off horn growth entirely, and they mature looking a lot like females.[3] Small, hornless males maneuver inside tunnels better than the bigger, horned males, in part because they don't have horns to get in the way.[4] Thanks to work by John Hunt, Joe Tomkins, and Leigh Simmons at the University of Western Australia, we also now know that the tiny males in many of these "dimorphic" dung beetle species are highly specialized sneaks. Quicker to mate and quicker to transfer sperm, they also have relatively bigger testes, and much more sperm.[5] They may not be able to mate as often as the weapon-wielding guarding males, but they make the most of the opportunities they get. Fighting wasn't working, so these little guys switched to plan B.

• • •

When only a few dominant males monopolize access to reproduction, there's strong incentive for the rest of the males to break the rules. If you can't win the game the normal way, cheat. Sneaky males are everywhere, lurking in populations of nearly every animal species.[6] Bighorn sheep rams guard harems on sheer slopes high in the Rocky Mountains. The largest and oldest males have by far the biggest horns, and these males consistently win ownership of the harems. Yet as many as 40 percent of the lambs end up sired by smaller males.[7] Called "coursers," these sneaky males charge into a territory for a few seconds at a time, forcing quick copulations with females before getting hammered by the dominant males.

Male sunfish and salmon guard cleared patches of sand where females come to lay eggs. Females choose large, attractive males with the best territories, sidling up next to them so they can shower sperm over her eggs. Tiny males have no chance of defending a territory or being picked by a female, so they dart in surreptitiously and squirt clouds of sperm into the mix.[8]

The biggest male ruffs, tall shorebirds that wade through marshes in Europe and Asia, guard territories from which they court, resplendent in gorgeous manes of fluffy black and chestnut feathers. Females consistently pick the largest, brightest males as mates, leaving few options for the rest. But smaller males cheat in two ways. One type of male dispenses with the black and chestnut plumage and dons a mane of white instead. These males are satellites, skirting around the edges of male territories and attempting to intercept females as they approach the dominant males.[9] The dominant males tolerate them, to an extent, because females appear to be drawn to territories with both dominant and satellite males.

Satellite males are obvious, decked out in white; but a third type of male mingles in the territories, too, and these males are tough to spot—so tough, in fact, that their existence in the mix was not discovered until decades after studies on ruffs began.[10] Called "faeders," these males look and act exactly like females. As a result, they're able to encroach onto the best territories unnoticed, edging up to females right in front of the territorial male.[11]

Sneaky males mimic females in a diversity of species. Marine isopod crustaceans—swimming pill bugs, for lack of a better term—guard hollow cavities in palm-sized sponges where females visit to feed and mate.[12] The biggest males wield fearsome claspers: tongs they use to grapple with rival males. Males with the longest tongs win, and these males succeed in guarding the best sponges. But other males work their way into the sponges, too. Medium-sized males dispense with weapons and look exactly like females. Like ruffs, female-mimic isopods are able to hang out inside sponges undetected by guarding males.[13]

Australian cuttlefish have the cleverest female mimics of all. These marine mollusks have superb color vision. Masters of camouflage, they're able to change the color of their skin in seconds so that it blends seamlessly with their backgrounds. For most of the year they live invisible and alone but, during a brief mating season, hundreds converge, and males begin to display, changing from dull, cryptic colors to a beautiful rainbow of greens, blues, and purples.

There may be as many as eleven males for every breeding female, so competition is fierce, and females approach the biggest, most colorful males.[14] Once a female chooses, the pair will attempt to swim to the periphery of the group to spawn. But sneaky males accost them in the process. The fact that they can change color so fast gives these mollusks a number of options. Sometimes little males charge in bright and strong, courting the females outright while guarding males are distracted in battles. Other times, they slide in looking like rocks, blending with the ocean floor as they glide up to the pair. Often, one will approach looking like a female, so that he can swim all the way up beside the dominant male unmolested. When the sneaky male has slipped in between the dominant male and the female he's mating with, working himself right beside the female, he will flash on his bright courtship display. However, he'll activate his display on only one side of his body—the side facing the female. He keeps the side visible to the dominant male cloaked and femalelike.[15]

• • •

Cheating and sneaking are as pervasive in human populations as they are in other animals, and they can stall even the largest armies.

"Irregular," or "guerrilla" warfare tactics date to at least the sixth century BCE, with the writings of Sun Tzu's *Art of War*.[16] The basic idea is simple: if a population faces invasion by an overwhelmingly superior army of conventional forces, break the convention. Don't fight them on their terms. By hiding out in local terrain and attacking only in swift, covert strikes, a smaller force can inflict irritating and demoralizing damage to a larger force. They'll never defeat the larger army outright, but they don't have to. Sneak forces "win" simply by staying in the game—by surviving, and by slowly draining the will to fight from the larger force.[17] Because irregular forces refuse to confront the larger army in traditional battle, they are all but impossible to expunge, and in the face of such secretive strikes, the bulk and strength of the conventional force actually becomes a liability.

The American Revolution would not have been named as such if the larger army had won. American rebel forces avoided direct, open-field battles with the better-trained and better-organized British army, opting instead for fleeting skirmishes, picking off troops as they marched or attacking when the larger army passed through choke points—narrow passages or river crossings that spread the army out, diminishing their numerical advantage.[18] Similar tactics were used against U.S. forces in Vietnam and the Soviet Army in Afghanistan. Today, U.S. forces grapple daily with sneak attacks by insurgents in Iraq and Afghanistan.

Guerrilla forces are "sneaks" for several reasons. In addition to breaking conventional rules of engagement, they use stealth and concealment to get close to the enemy, ideally approaching unnoticed until the moment of attack. They also "hide" in the sense that they rarely wear military uniforms. By blending in with the civilian population, guerrilla forces make it difficult for the invading army to tell friend from foe, a "lose-lose" situation for the army; erring on the cautious side risks death from attacks, while overreacting kills noncombatants and undermines political support for the campaign.[19]

The biggest, most expensive weapon technologies sometimes succumb to sneak attacks. Foot soldiers don't stand a chance in a direct confrontation with a tank, but a grenade or Molotov cocktail slipped into a hatch can change the odds considerably. Land mines and improvised explosive devices (IEDs) cheat, in the sense that they avoid a direct confrontation and lurk, instead, in the shadows, hidden in rubble or under the soil. These tiny weapons can disable multimillion-dollar tanks and armored vehicles, and subsurface mines can sink billion-dollar battleships. In October 2000, a tiny craft sailed alongside the USS *Cole*, a five-hundred-foot-long, $900 million guided-missile destroyer. The craft appeared benign, but was, in fact, laden with explosives, blowing a forty-foot hole in the hull of the destroyer, killing seventeen, wounding thirty-nine, and inflicting $150 million in damages.[20]

The most dangerous form of sneak attack, at least against our modern forces, may also be the least appreciated. Cyberattacks don't sound very scary, and it's tough to imagine how they could threaten our security beyond the occasional hassle of a usurped credit card password or identity theft. But hacking may prove to be this country's greatest danger, capable of crippling the entirety of our military forces.

Over the past few decades, military technology has become increasingly computerized. Everything from the guidance systems on our missiles to the navigation and handling of our submarines, aircraft carriers, and aircraft relies entirely on high-tech software. Modern aircraft fly at speeds and in maneuvers that push the limits of human capabilities, and the most modern craft are impossible to fly without the aid of computer-enhanced flight controls.[21] Targeting, flight-control, navigation, and even command and control, all depend critically on sophisticated electronics and software.

Hackers sneak into our military mainframes, bypassing firewalls and surreptitiously inserting foreign code. Between 2003 and 2006, for example, Chinese hackers launched a series of coordinated cyberattacks against U.S. defense and aerospace installations.[22] Before they were finally detected and the wormholes filled, these "Titan Rain" attacks pulled masses of sensitive military data from the U.S. Department of Defense, the Pentagon, NASA, Los Alamos Laboratories, Boeing, Raytheon, and other sources. Titan Rain demonstrated with startling clarity how China could use cyberwarfare as an asymmetric tactic to undermine conventional forces of an adversary.[23]

In 2013, it became clear that China was at it again, this time worming into the control systems for an alarming number of our most advanced weapons, including, among others, the F-35 joint strike fighter and the V-22 Osprey tilt-rotor aircraft, the Terminal High Altitude Area Defense missile system, the Patriot Advanced Capability antimissile system, the Aegis Ballistic Missile Defense System, and even our Global Hawk unarmed aerial vehicle system.[24] The fact that these crucial weapons were compromised is scary enough, but the truly terrifying part of these events was the realization that the Chinese were not simply pirating information. It now appears their plan was to inject code that would, when activated, give them complete control of our systems.[25]

"Zero-day" attacks, as these are called, are the hacker's most difficult and dangerous weapons, codes so deeply embedded that they lurk invisibly until the day they're needed, exploiting vulnerabilities that even the software makers are unaware of.[26] Had they not been discovered, codes inserted during the 2013 sneak cyberattack could have utterly incapacitated the most expensive and advanced weapons in the history of mankind, possibly even turning them against us.

• • •

Coursers, squirters, satellites, and female mimics—there are many ways to cheat. For animals, this means that in addition to traditional confrontations with armed rivals, dominant males now face the more insidious threats of males breaking the rules. Similarly, guerrilla forces, land mines, IEDs, and cyberhackers all can undermine the effectiveness of conventional military forces. As long as the impacts of these cheaters are relatively minor, not much changes. When cheaters begin to get too effective, however, they can end an arms race.

11. End of the Race

During the height of the Middle Ages, the strength, weight, and cost of armor reached unprecedented extremes. In staged tournaments where knights fought similarly armed rivals in one-on-one jousts, bigger was better, and knights with the best equipment usually prevailed.[1] The advantages of armor held on traditional battlefields, too, where men-at-arms advanced to face opposing armies head-on. Suits of armor eventually got so heavy that men and horses became clumsy,[2] forcing formations to march straight ahead into battle without sudden—or, indeed, any—changes in direction.[3] But as long as they clashed with comparably armed opponents facing similar limitations, the best-trained, best-equipped men in the best armor still won, and the protection offered by armor justified its excessive cost. In fact, the knights were so well protected that most battles of this sort were settled with surprisingly few casualties. Loss of honor sufficed in lieu of death or dismemberment.[4]

New types of weapons changed all of this, including the crossbow and, later, the English longbow.[5] Like sneaky males, these new technologies "cheated," breaking the rules of engagement in ways that eroded the benefits of expensive armor. Prior to the invention of the crossbow a knight could ride into battle immune to all but other knights. He could sweep through a field of peasants, slashing from on high as he charged, with sheets of mail and plate armor, shield, and helmet protecting him from anything hurled up by foot soldiers below. Knights alone had this advantage. They towered over the ill-protected, shoddily armed peasant fighters, dismissing them as a waste of effort and seeking out knights of status from the opposing ranks instead.[6]

Armed with crossbows, however, ordinary farmers could shoot down the best-trained, best-armed knights of the day. Suddenly, sitting astride a horse was a problem, rather than a tactical advantage, because mounted knights made easy targets. Bolts fired up at them could slip beneath armor plates—into the armpits, for example—and direct hits penetrated regardless. Horses could be toppled, too, bringing the mass of muscle and metal crashing to the ground so that a knight lay underfoot, as helpless as an overturned turtle.[7]

Crossbows and longbows broke every rule of engagement that mattered for the evolution of armor. Unlike armor, these weapons were cheap and relatively easy to use. They were not the exclusive purview of the wealthy elite. They didn't require a lifetime of training, and as a result, they were not accurate reflectors of status or class.[8] Most important of all, crossbows and longbows changed the structure of military engagements. Armies adopted new tactics to incorporate these weapons, and battlefield match-ups between rival knights gradually disappeared.[9]

At the battle of Crécy, for example, English forces positioned themselves in terrain designed to concentrate fire from their archers onto the advancing French.[10] Edward III chose a flat agricultural field flanked by forest and other natural obstacles, and ordered his men to dismount and wait, rather than advance on horseback. Three divisions, each containing one thousand men in armor (called "men-at-arms"), were arranged six deep in rows flanked on either side by roughly five thousand archers. A thousand men-at-arms were held in reserve as cavalry, ready to pursue the French in the aftermath.

The combined strength of the English was approximately twenty thousand, of which four thousand were men-at-arms. The French, on the other hand, advanced with three times this number, including twelve thousand men-at-arms who stayed on their horses.[11] The first French to advance were six thousand crossbowmen, mercenaries hired for the occasion. Behind them advanced the ranks of men-at-arms. The army marched to within 150 yards of the English, and the crossbowmen fired their bolts, but most fell short of the emplaced army. So they advanced again, only to be met by a shower of longbow arrows, shattering their ranks and triggering panic. Impatient to meet their equals in battle, the French cavalry charged forward anyway, plowing through and over their scrambling mercenary crossbowmen only to find themselves trapped in a deadly crossfire. Horses stumbled and collided, tripped over dying men, or were shot, and knights toppled to the ground. The few who reached the English line were routed, but the French kept advancing. More than a dozen waves surged forward, all tangling in the jumbled piles of bodies, trapped in a slaughter. When the French finally gave up, they left more than fifteen thousand dead. The English lost only two hundred.[12]

Seventy years later, Agincourt ended the same way. Although the

Unhorsed knights were vulnerable to weapons that broke traditional rules of engagement, like the English longbow and, later, muskets.

French outnumbered the English by five to one, their numerical advantage vanished in showers of metal-tipped arrows, as piles of bodies tripped up knights in the fray.[13] Unhorsed and stuck in the mud, fully armored knights proved easy prey. The peasants they detested and ignored in traditional battles shot them point-blank with arrows, and what had been the ultimate protection before—magnificent armor untouchable by the masses—become a deadly liability. An inexpensive new weapon made elaborate armor obsolete.[14]

• • •

No arms race lasts forever. As weapons get bigger they also get dramatically more expensive. Eventually, populations reach a new balance where the now-higher costs neutralize the reproductive benefits. Bigger stops being better, and the arms race stalls. Populations stabilize, hovering at the new weapon size. How big the weapons are at this stalling point depends on where the balance is finally reached; animal weapons under strong sexual selection, for example, may attain astonishing proportions before costs catch up and place the process in check.

Populations in such a balance can persist for long periods of time, holding fast with massive structures. Were you to examine one of these populations and measure the strength of selection on weapons, you might find it is weak at best, or absent.[15] When people look at species with huge weapons the last thing they expect to find is weak selection for weapons, but it shouldn't be surprising. In theory, every population reaches such a balance eventually, and, once they get there, they should stay there. Considering how rapidly weapons can evolve and how many species appear to have had them for millions or even tens of millions of years, it's only logical to assume that many have reached their balancing point, and their arms races have stalled. The tape on their tug-of-war rope holds fast, powerful forces canceling each other out as they strain in opposite directions.

Cheating males work just like costs, offsetting the reproductive advantages of large weapons. For males to reap the full benefits of their weapons, success in battle must translate into success siring offspring. In a perfect world, victorious males sire all the offspring of females they defend. In reality, sneaky males surreptitiously mate with guarded females, stealing fertilizations and undermining the effectiveness of the guarding tactic.

If a sneaky male dung beetle sires one-fourth of the guarded female's offspring, for example, this slashes the payoffs to the dominant male by the same amount. He still pays to produce his horns, and he still pays to use them, fighting constantly to keep intruders at bay, but now the reward he earns is only 75 percent of what it otherwise could have been.[16]

Averaged across the population, some fraction of male success is always likely to go to sneaky males. After all, cheating tactics are ubiquitous in the animal world; nearly every population has them. As long as sneaky males steal only small amounts of sired offspring, their effect on weapon evolution is likely to be minimal. However, when sneaks start doing well, they may erode the payoffs to fighting males substantially. Together with traditional costs, reproductive success lost to cheats can put the brakes on continued weapon evolution, helping define the point where populations begin to stabilize. In fact, if cheaters start doing *too* well, they may erode payoffs to weapons so drastically that the direction of selection reverses; big weapons become a liability. Instead of stalling, these races collapse.

• • •

Once the rewards for big weapons plummet, selection starts to favor rapid reductions in weapon size. In theory, if populations with costly weapons fail to lose them fast enough, they may even suffer extinction. We will never know exactly what happened to the majestic Irish elk, for example, but the latest attempts to model costs of antler growth indicate that antlers had reached a very expensive extreme, in which even the most successful males barely managed to recoup their calcium and phosphorus losses before the onset of winter.[17] This, combined with climate reconstructions showing a dramatic decrease in the mineral contents of the plants the elk foraged, raises the possibility that a change in habitat pushed populations into a place where they could no longer afford to pay the price of their extravagant weapons.

All we know for sure is that this drop in food quality coincides with their extinction.[18]

More often, I suspect, populations persist but their weapons disappear. When a few species of stalk-eyed fly stopped roosting on hanging rootlets in Malaysian streams, two of the three arms race ingredients vanished; females stopped gathering in harems on hanging threads (they were no longer economically defensible), and fights stopped being duels. With two of the race-stimulating ingredients gone, the flies' arms races collapsed. They lost their extreme weapons.[19]

When one species of stag beetle dispensed with battles over sap flows, scrambling instead along the broad surfaces of the insides of hollow trees, resources stopped being economically defensible, and mandible sizes shrank.[20] Similarly, males in three additional stag beetle species started forming stable, long-term pair bonds with single females and helping them raise their young—causing two of the three arms race ingredients to disappear. Now, not only were males not battling over localized sap sites, they weren't even battling at all. The reproductive turnaround times of males were so similar to those of females that there was almost no competition. Today, their mandibles are tiny.[21]

Environments change, and there is no a priori reason to presume that conditions conducive to arms races will last indefinitely. In fact, most clades of heavily armed species show ample evidence of weapon loss. Reconstructions of the histories of these groups reveal a multitude of evolutionary gains and losses, suggesting that weapon evolution is a dynamic, even cyclic, process of escalation and collapse. When my colleagues and I explored patterns of weapon evolution in a sample of fifty dung beetle species, for example, we found new horns had been gained fifteen separate times—the cauldron of preconditions sparked one arms race after another. But horns had disappeared, too. Nine times the race collapsed, and species lost their weapons.[22] Similar studies of horns in antelope show both increases and decreases in weapon size.[23] Arms races are like houses of cards, magnificent and fragile.

• • •

Whenever a small number of wealthy states invests in exorbitant weapons that nobody else can afford, someone, somewhere, is sure to come up with a cheap way to bring them to their knees, and, every so often, the smallest of weapons manages to topple a Leviathan. Fire ships brought panic and disorder to naval fleets as far back as the fifth century BCE, when the Syracusans took an old merchant vessel, packed it with pitch and pine, set it aflame, and released it to drift downwind into the Athenian navy.[24]

Two millennia later, fire ships were still used in essentially the same way. Wooden hulls and stores of gunpowder necessary for loading cannons made sailing ships of the line extremely vulnerable to fire.[25] Small, disposable vessels packed with flammable materials or explosives could wreak havoc, especially if they were set adrift to blow into a full line of battle. A mile-long string of warships, all sailing in strict formation one behind the other, made an easy target even for a drifting vessel, and the fire ships were much too small for crews to hit readily with cannon fire, permitting them to slip dangerously close to the giant vessels.

Although fire ships only rarely sank a ship of the line, they often forced fleets to break formation in ways advantageous to the instigators.[26] In 1588, for example, the English sent eight fire ships drifting into the anchored fleet of the Spanish Armada, 140 warships clustered together in darkness off the Calais shore. The English could not afford to let the Armada stay moored, since reinforcements were likely on the way, and they needed to force them out into the open. The Spanish saw the fire ships coming—indeed they had been expecting them—and were able to catch and divert two away from the fleet. But the remaining six slipped past the outer defenses and dispersed the Armada. None of the Spanish ships caught fire that night, but the fleet was forced to weigh anchor in darkness. By dawn the next morning, the English had engaged them in battle.[27]

It was a new style of gun, rather than sneaky fire ships, that brought the age of the sailing battleship to its close, however. Advances in artillery, particularly rifled barrels and exploding shells, rendered these warships obsolete.[28] For the preceding three hundred years, naval guns had all been smooth-barreled cannons firing solid iron balls.[29] Guns and balls got bigger, but the technology remained the same. By the 1850s, however, spiral grooves were cast into the bores, or barrels, of cannons, causing projectiles to spin. "Rifle-bore" cannons struck targets with far greater accuracy and at longer ranges. Around the same time, cannonballs were replaced by pointed shells filled with explosives, utterly changing the type of damage inflicted in naval battle.

Solid iron balls smashed holes into hulls, toppled masts, and impaled crews with flying shards of timber. But it took a lot of iron balls to sink a ship. For this reason, bigger ships were better in a brawl. They packed more cannons and, therefore, fired a bigger broadside. Thick wooden hulls and bigger broadsides rendered the largest ships invulnerable to all but an equally matched rival, precisely the conditions conducive to an arms race.

Explosive shells cheated. They blew metal shrapnel everywhere, rupturing hulls beneath the waterline and triggering catastrophic fires.[30] Wooden hulls simply could not withstand the destructive impact of exploding artillery. Just a single direct hit could sink a ship. Furthermore, since these new guns could be mounted on small, inexpensive vessels, they undermined the traditional rules of engagement. All but overnight the most impressive warships became bulky targets.

The solution was metal armor shielding the sides of ships, but steel hulls weighed too much to power with sails.[31] The arms race between sailing vessels was over. Navies were at an impasse until steam-powered screw propellers released ships from the weight constraint, launching a new arms race between ships powered by propellers instead of wind.[32]

Epitomized by HMS *Dreadnought*, a sleek and imposing new warship took center stage. Sheathed in armor and sporting large, rifled cannons on rotatable turrets, these new "ironclads" killed from afar. Once new aiming techniques had been perfected, ironclad battleships could sink opponents from several miles away.[33] Fights no longer unfolded yardarm-to-yardarm, but duels between these great ships persisted nevertheless, and the nature of these new guns created an environment where bigger was better—more than enough to trigger a race.

Armor resisted small-caliber shells, which gave ships with bigger guns an advantage. Bigger guns, in turn, selected for thicker armor, and both required bigger ships to carry them. At the same time, revolutions in engineering were advancing the power provided by steam propulsion. The race for speed was on. Navies rushed to develop ships that were bigger, faster, more heavily armored, and sporting larger guns than ships in rival navies. The burst of battleship construction that followed has been described as one of the most rapid and prolific arms races of all time.[34]

At first Britain, France, Germany, Russia, Italy, the United States, and Japan all launched into massive shipbuilding campaigns but, by the beginning of the twentieth century, the sheer cost and staggering sizes of the fleets caused the race to coalesce around just two naval superpowers, Britain and Germany.[35] By the start of WWI, dreadnoughts had evolved into larger, faster "superdreadnoughts," and both countries managed to assemble dozens of these outrageously expensive ships.[36]

As impressive as these new navies appeared, they were threatened from the start by cheaters. Fire ships had evolved into torpedo ships—small, fast boats able to slip up close to a battleship and release motorized torpedoes. Bulky battleships could not maneuver fast enough to evade these little pests, so navies created destroyers—small, specialized warships designed to intercept and sink torpedo boats.[37] Soon, destroyers themselves carried torpedoes, making them useful for offense as well as defense.[38] Just as ant colonies produce both big-headed, powerful soldiers and smaller, mobile workers, fleets now spawned big ships and little ships designed around specialized tasks.

But the sneakers kept getting better. Soon it was possible to carry torpedoes in underwater ships, firing them without ever breaking the surface. The ultimate sneaks, submarines, could creep up to the biggest battleships and sink them from below,[39] undermining the prestige and tactical effectiveness of a navy's most powerful weapons. Like peasant farmers with crossbows, submarines shattered the traditional rules of engagement. Battleships became ungainly targets; their greatest strength, a liability. The only way to use them was to surround them with other ships, gobbling up fleet strength and limiting movement since each behemoth now had to be escorted by a gaggle of destroyers and escorts running interference.

The Germans apparently realized they would never catch up with Britain's naval construction frenzy, so they secretly shunted funds from battleships to submarines, assembling a fleet of stealth U-boats.[40] Ironically, Germany used its U-boats to greatest effect not by targeting and sinking naval battleships, which were now protected by squadrons of destroyers, but by sinking merchant ships crossing the Atlantic unprotected.[41] There was no way the British (or any) navy could protect all of these merchant ships. They became easy targets for the submarines, crippling transport of war materials and personnel and undermining the Allied war effort.

Submarines, in turn, spawned still another form of sneaking. The British cloaked military ships to look just like helpless merchant vessels. "Q-ships," as they were called, were one of the most closely guarded secrets of WWI.[42] The trick was to lure submarines in close and induce them to surface. Subs carried only a limited number of torpedoes, so if the merchant vessel appeared helpless enough, a submarine might surface to sink the ship using deck-mounted guns, saving precious torpedoes for a later target. Sometimes Q-ships would feign

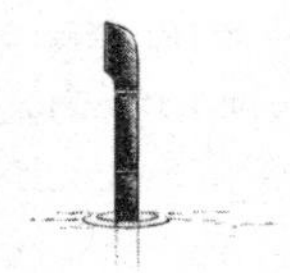

Submarines were the ultimate "sneaks" in naval warfare. Small and invisible, they could sink even the largest of battleships.

sinking, releasing smoke while crew members abandoned ship in lifeboats, enticing the U-boat to surface to finish the job. The moment the sub breached, remaining crew on the Q-ship dropped panels along the hull, revealing hidden deck guns and opening fire. The Q-ship double sneak was a bold and clever strategy, but it proved exceedingly dangerous and, ultimately, cost-ineffective. Q-ships sank fourteen German submarines during the war, but twice that many Q-ships were lost in the process.[43]

By 1914, it was clear even to the British that despite their grandeur, battleships were not going to prove decisive in battles. They'd been designed to fight other battleships, but such engagements almost never materialized and, as deterrents, they had become obsolete.[44] Although battleships remained a part of naval fleets for many years afterward, their prestigious role as monarch of the seas was over, and the few that remained in service ended up supporting a newer and better weapon, the aircraft carrier.[45]

In the end, the fate of all weapons comes back to benefits and costs. Early in arms races the payoffs for big weapons may soar. But circumstances change. Costs climb and cheaters invade, gouging profits until big weapons reach a point that's unsustainable: the rewards no longer justify the expense. From that point forward, they're just a liability.

• • •

The sunset was beautiful, as it nearly always was on Barro Colorado Island. Colorful parrots squawked as they converged from all directions, piling into a communal roosting tree at the edge of the water. A keel-billed toucan crossed the clearing in front of us with a soft *whoosh.* It was February 1992. I'd completed my stint in the Panama forest and was preparing to return to normal life as a graduate student. All the beetles were measured; the ant farms disassembled; and thousands of plastic tubes sat boxed and secure in deep storage. My lab was clean and my bags nearly packed. For almost two years I'd lived and worked at the Smithsonian research station on BCI, hanging with a crowd of biologists all working diligently to understand the details of life. Now it was time to go home.

Several of us relaxed on a porch overlooking the canal, beads of precipitation dripping from our ice-cold bottles of beer. Big ships were a regular sight, as boats traversed from the Atlantic to the Pacific or vice versa. Most were boring box boats, cargo carriers stacked to the brim with metal crates. Every now and then a cruise ship would appear. But the most exciting by far were the warships. This evening an entire fleet of the U.S. Navy glided quietly by—destroyers, cruisers, and, best of all, a battleship—a slate-gray monstrosity, awe-inspiring with its gigantic guns, bristling with antennae, radar, and satellite dishes. U.S. battleships were the largest vessels able to fit into the canal, we were told. At almost nine hundred feet long and just over one hundred feet wide, *Iowa*-class battleships had only *eleven inches* to spare when they squeezed into the locks.

I'll remember that night for the rest of my life, because I woke to a shattering explosion. In a flash of blinding light I was thrown six feet from my bed, hitting the wall and then crumpling to the floor. I sat there, shaken and aching in the sudden darkness, sure we were at war; my first thoughts were of the battleship. Why would it fire at the field station? This was not quite as preposterous as it might sound. After all, just twenty-five months earlier the United States had invaded Panama, ousting Manuel Noriega in Operation Just Cause. Bullet holes still marred the walls of buildings in Gamboa a mere ten miles away.

Of course, we were not at war; the battleship was just a coincidence. Bolted to the other side of my skimpy dorm room wall was a hundred-foot-tall radio tower, and although it was supposed to be grounded, lightning had struck it that night with enough charge to zap me to the floor—one of the reasons, no doubt, that the building was dismantled a few years later and the clearing returned to the forest.

I didn't appreciate it at the time, but that glorious ship, cruising proudly into the fading light, was the USS *Missouri*, the last active battleship in the world. She was decommissioned a month later that same year. I had witnessed the final voyage of the very last battleship of them all.

PART IV

PARALLELS

The first three sections of this book focus on animal weapons, plucking snippets from military history as needed to make a point or to illustrate a parallel. Just how far do these parallels extend? The final chapters delve more deeply and completely into humanity's greatest arms races, revealing startling similarities with animals, as well as important differences.

12. Castles of Sand and Stone

African army ants, like their Central and South American relatives, are ferocious predators. Jaws of the largest soldiers can snap through a pencil, but the true force of these ants lies in their numbers. A raiding swarm twenty million strong pours forth from its nest, advancing in a column through the bush. The leading wave of ants is so thick that it drives everything that can to move out of its way. Anything that doesn't faces a horrific death. Animals overtaken by a swarm are sliced into ant-size chunks, hoisted by workers, and carted piecemeal back to the nest. Streams of workers carry the carnage of a raid along ant superhighways—cleared trackways coalescing like giant, exposed blood vessels pumping blood back to a heart. As they race homeward with their loads, the workers are protected by walls of soldiers flanking the sides of the trackway and, in some places, covering it entirely with a latticework of their intertwined bodies.[1]

The Kenyan Maasai love "siafu," as they call these ants, since they scour through houses cleaning out cockroaches, detritus, other ants, and even rats. The Maasai also use siafu as emergency sutures, in precisely the same way that I did years ago in Belize. But African army ants have a dark side. Occasionally, livestock get trapped and, when prevented from escaping, they, too, are devoured. Coops full of chickens, tethered goats, and cows can be stripped to bones in just a few hours. Sometimes even elderly or drunk people fall victim. Eighteenth-century explorers describe the use of army ants as a cruel form of execution, binding criminals to a stake in the path of an advancing swarm.[2] But the most tragic human fatalities involve infants left unattended in their cribs. Ants march into a nursery through an open window, swarm up the sides of the crib and pour into the mouth and lungs, suffocating the baby as its flesh is stripped away. As many as twenty infants every year still succumb to siafu.

In January 2005, Caspar Schöning, now a biologist at Freie Universität Berlin, watched an army ant raid unfold at the edge of a meadow behind the field station where he was staying. Schöning had completed his doctoral studies the year before, focusing on the behavior of army ant swarms; he was in Nigeria with renowned nature photographer Mark Moffett to photograph a raid. What they captured was highly unusual. At first the ants carted back pieces of beetle and cricket—typical fare—along with shards of spider and moth, the pieces bobbing back and forth as the ants marched. But then a string of soft, white grubs began to appear. The ants were carrying termites. On and on the ants came, carting pale termite bodies back to the nest. Thousands and thousands of them streamed past. Big-headed soldier termites flowed by in pieces. Heads carried by one worker, legs by others, abdomens by still others. Soon the termite brood began to appear, tens of thousands of eggs and larvae shuttled along an endless conveyor of running ants. Schöning and Moffett estimate that more than half a million termites were snatched that night, including the colony's offspring.[3]

That the army ants had successfully plundered a termite colony was surprising, because siafu essentially never eat termites.[4] In fact, Schöning and Moffett's published account of this raid remains to this day the only documented example of its kind. Army ants are not exactly picky about their diet choices—they'll eat anything from spiders to cows, and everything in between—and termite colonies litter the landscape, making them one of the most abundant sources of prey in the area. Termites are soft and plump, loaded with lipids, protein, and carbohydrates, and, except for the soldiers, they are utterly defenseless. When they're exposed aboveground, just about everything eats them. All of which makes the omission of termites from siafu diet striking. The secret to termite safety: fortresses.

Termite mounds are magnificent structures.[5] The species raided that day by siafu, *Macrotermes subhyalinus*, erects castles of sand and mud towering ten feet above the ground—two thousand times taller than the termites themselves. The conical base of a mound may be four feet across, but it constricts as it rises to a central spire, and the top eight feet jut skyward like a stovepipe. On the outside is a solid wall. Meticulously erected by millions of workers, the wall is built from grains of sand mixed with excrement and spit to form a cement. Baked by the sun, the outer wall is as durable as kiln-fired brick. It takes a sledgehammer or an ax to breach the surface, and it's not uncommon for sparks to fly with each strike.

If you manage to smash through this wall, you'll find empty space on the other side. No termites yet. Inside the outer wall sits a second wall. This inner wall is separated from the first wall by the insect equivalent of a castle's moat—a no-man's-land. This air-filled space forms a six-inch gap, and only after breaching the second wall will you get to the colony proper. The termites themselves enter or exit through tiny little gates, a half dozen termite-sized tubes that extend through the no-man's-land to the outside world. These tubular gateways have rock-solid walls, and they're guarded by masses of big-headed soldiers.

At the core of the fortress a city of insects bustles.[6] Millions of workers flow back and forth between dozens of coconut-sized chambers. Like coconuts, each chamber is encased in a rigid protective shell. They stack together in the soil in a bundle connected by a maze of tunnels and sheltered from the rest of the world by the concentric rings of the outer walls. Some of the chambers house food for the colony, a fungus that the termites cultivate in the cool darkness belowground. Others are nurseries packed with eggs or larvae. At the heart of the complex lies the most protected chamber of all. Like the keep of a medieval castle, the chamber housing the queen is set apart from the rest, accessed only by a single opening so small that the queen herself could not possibly fit through.

Termite queens are as specialized for reproduction as the soldiers are for battle. Egg-laying machines, queen termites have bloated, pulsating abdomens thicker than my thumb, and they lay thousands of eggs each day. Minuscule workers wait beside the queen, plucking new eggs from her body as they emerge and shuttling them off to nurseries. The queen is so obese that she literally cannot move. Completely dependent on the workers to feed and tend to her, and irreversibly ensconced in her inner ward, she'll live for ten years or more without ever leaving the protection of the royal chamber.

Millions of busy workers and balls of growing fungus consume huge amounts of oxygen and release carbon dioxide. They also generate heat, and both must be dissipated from the colony for it to thrive. Remarkably, the architecture of termite mounds accomplishes this as well. Although hard as a rock and impenetrable to other insects, the walls of termite mounds are infused with hundreds of millions of microscopic pores. The walls breathe, allowing oxygen to enter and carbon dioxide to escape. Wind passing over the chimney—or even directly through the walls—pulls stale air out of the colony and replaces it with fresh, oxygenated air. The double-walled design also works like an insulator, keeping the colony inside at a constant and cool temperature.[7]

Termite mounds provide protection, storage, and temperature control, but their primary function is protection. Impenetrable walls shelter each colony from attack. Short of a Godzilla-like monster ripping the top off—which for Schöning's colony meant the claws of an aardvark—the only way into a mound is through one of the tiny, heavily protected gates. The heads of the termite soldiers who guard these gates are so big that they can barely walk, and eyes and other vulnerable organs are long gone. Giant jaws agape, soldiers waddle into the fray, lunging in droves at any insects that attempt to pass. Snapping jaws lock onto legs, sever heads from bodies, and snip off antennae, reducing invaders to jumbles of parts. At the same time, worker termites seal off the gates from within. The moment attacking ants are detected, an alarm signal spreads and workers flock to the gates. Like lowering wrought-iron portcullises to block entry to a castle, these tiny masons cement sand and mud into each tunnel gate until it is completely plugged. Only later will they reopen the gates, long after the raiding ants have marched on.

• • •

Throughout history we've built walls around our cities for precisely the same reason the termites do. In military-speak, walls act like a "force multiplier," permitting a small number of defenders to hold out against a much larger invading army. The moment people settled into stable populations they became vulnerable. Agrarian societies could not wander the way earlier peoples did, since their crops were rooted in place. Surpluses of food could be stored for later distribution during harder times, enabling populations to grow. But stored foods could also be stolen. Like termites, people in early civilizations found themselves in a tight spot. They needed to find ways to protect their food and their brood from nomadic raiders. The earliest remnants of settled towns, scattered along the banks of rivers such as the Tigris, Euphrates, and Nile, all show evidence of walled fortifications, some dating to more than 5500 BCE.[8]

The first defenses appear to have been ditches fronted by wooden palisades but, by 3500 BCE, cities were surrounding themselves with walls built from mud brick and stone.[9] Askut, Semna, Uruk, and Jericho all had populations numbering in the tens of thousands. Each was protected by tall brick or stone walls and was entered only through heavily fortified gates.[10] By 1500 BCE, cities were experimenting with double walls, providing two successive lines of defense. Walls were topped with walkways and protected by battlements, alternating plates and slits, which gave cover to archers shooting down onto attackers. Towers bulged out from the walls at regular intervals, and from them defenders could fire at the flanks, or even the backs, of invading troops attempting to scale or breach the wall. Battlement walkways jutted several feet out from the face of the wall, looming over the space below like a balcony. Holes in the floors of these protruding battlements let defenders drop boulders onto the heads of attackers, pour boiling oil, or even empty chamber pots and garbage onto the faces below.[11]

The Judean city Lachish exemplified this style of ancient fortification.

Perched atop a rock butte, the city sat two hundred feet above the plain below, replete with houses for its eight thousand residents, markets, synagogue, and an eight-hundred-foot-deep stone-lined well.[12] A narrow road led up the side of the butte, slipping in between two enormous stone towers—the gatehouse. Each gatehouse tower loomed fifty feet high. Surrounded by jutting battlements with archer slits and floor holes, gatehouses promised intense flanking fire from all sides and above, a deadly crossfire to anyone attempting to force his way through the gate. Inside this gate was another, and then another. All told, the Lachish gatehouses comprised six successive "kill zones"—exposed chambers through which intruders needed to pass and in which they would have been exposed to arrows, oil, and boulders from all sides.[13]

Wrapped all the way around the outcrop were two tall walls. The outer wall, forty feet high and ten feet thick, circled the city roughly halfway up the escarpment, making climbing the hill virtually impossible or at least dangerously impractical. At the very top of the outcrop rose another city wall, taller and thicker than the first, and both walls were capped with protruding balcony-like battlements.[14] Although by no means unusual for the time, the fortifications of Lachish were impressive. Combined with the elevated and inaccessible location of the city atop the rock hill, they made a forbidding barrier to any invading army.

But the Assyrians weren't just any army—they were unmatched in their day—and they were ready for this challenge. Like siafu, Assyrian society revolved around warfare, and they had huge, well-trained, professional standing armies at their disposal. No one dared face Assyrian archers and chariots in open-field battle, which meant they'd had lots of practice overtaking fortified towns. When the Assyrians advanced on Lachish in 701 BCE, and settled into camp on the adjacent hill, they came prepared to breach its walls.

Assyrian siege strategy applied overwhelming force to many parts of the fortress at the same time. By spreading the defending troops thinly, forcing them to cover multiple attack points simultaneously, the Assyrians dramatically increased the odds that one of their assaults would succeed. But assaulting fortress walls required planning and preparation. The tools needed for successful siegecraft were so big they had to be constructed on-site, and this could require many months of labor. In addition to foot soldiers and charioteers, Assyrian armies brought with them thousands of engineers, trained builders skilled at erecting temporary fortifications around their encampments and at building the towers and tunnels needed to overthrow heavily defended fortresses.[15]

Assyrian siege towers stood three stories tall, permitting attackers to fight from the same height as the defenders. Wooden battlements lined the top floors, giving archers cover as they fired. Each tower was fronted with a drawbridge that could be dropped into place when the tower got close enough to a wall. Middle levels in the tower carried additional troops prepared to rush up ladders as soon as the drawbridge was lowered. The bottom floor housed a battering ram. A giant iron-tipped log hung suspended from the frame so that it could be swung back and forth into the fortress walls. Like a massive crowbar, a wedged ram could also be rocked side to side, prying stones from the wall once cracks started to form.[16]

The problem with siege towers was that they had to be pushed into place, and defenders positioned ditches, moats, and outer curtain walls to impede such progress. This meant that engineers often had to first drain and then fill in moats with rocks and soil, and then construct roadways leading up to the walls. In the case of Lachish, the problem wasn't a moat; it was cliffs, and the forty-foot wall that loomed halfway up it. The embankments beneath the wall were way too steep for siege towers, so the engineers built a ramp.

Stone by stone, the Assyrians assembled a massive causeway two hundred feet wide at its base, tapering as it rose to fifty feet wide when it finally abutted the city wall. When it was finished, the ramp climbed high above the plain below, reaching half way up the cliffs of the escarpment—high enough that the tops of the siege towers lined up evenly with the battlements of the outer city wall. The Assyrians used captured prisoners from nearby towns to pile the stones beneath a barrage of hostile fire, forcing the Lachish defenders to shoot their own countrymen to slow the incessant advance of the ramp.[17]

At the same time as the ramp was being constructed, engineers built five siege towers, each mounted on giant wooden wheels. They also manufactured dozens of tall ladders with hooks that could be thrown against the walls in haste, as soldiers rushed the perimeter. When all of the pieces were finally in place, the Assyrians launched their coordinated attack. Soldiers inside the bottom floors of the siege towers pushed their contraptions side by side up the ramp toward the city walls. The wooden-sided towers shielded soldiers from arrows, and water-soaked hides covered all exposed surfaces to protect against flame. Rows of archers marched alongside the climbing towers showering the defensive battlements with arrows, and each archer was accompanied by a shield bearer to keep his hands free for firing.[18]

When the wheeled siege towers reached the Lachish wall, other ladder-bearing soldiers launched additional attacks. Flanked by throngs of rapid-firing archers, foot soldiers with shields, pikes, and swords swiftly scaled the walls on all sides of the city, drawing defenders away from the main gates and the wall at the top of the ramp where the battering rams were smashing. The siege towers crashed through the defenses, and Assyrian soldiers poured into the city proper. The result was a savage and total destruction: city walls torn asunder, buildings demolished, and the city leaders skinned alive. Attackers impaled defending troops on stakes or blinded them with swords, and slaughtered thousands of the city's inhabitants. Those not killed on the spot were deported, marched by the Assyrians to far-off lands where they lived the rest of their lives as slaves.[19] Like the termite colony plundered by siafu, Lachish was gone.

• • •

The similarities between these two battles are striking: both involved a sedentary population defending itself against attack from a larger, invading army. Both cities were surrounded by rigid walls, which restricted entry to a small number of heavily fortified gates. And, in both cases, the fortifications ordinarily sufficed, protecting inhabitants against the majority of likely threats.

Most armies of Lachish's day lacked the engineers, supplies, and time necessary for waging a prolonged siege. To starve out a defended city, or breach its walls, required tens of thousands of soldiers camped for many months—sometimes years—in hostile territory a long way from home. The invading army had to protect itself during this process, so they often built a fortress around their camp, which required huge quantities of food and timber. Also, since they would be far from home for an extended period, invaders needed to leave the bulk of their armies behind. That is, their armies had to be so vast, and so well organized, that they could move ten thousand troops to a foreign land without leaving their home cities vulnerable. All told, the logistics involved with successful siege warfare vastly exceeded the capabilities of all but the richest of armies.[20] Most of the time, the only option available to an invading force was to storm the walls and rush the gatehouse. And in this situation, gatehouses were spectacularly effective.

Walling a city essentially converts it into a tunnel. By forcing intruders to enter through a narrow gate, a city removed the numerical advantage of invading troops. Regardless of how many soldiers waited outside a city's walls, only a few could enter at a time and, in the narrow confines of gatehouses, soldiers defended from protected positions—their flanks were shielded by slabs of stone; those of the attackers were not.[21]

Narrow gates work the same way for termites. The strength of siafu armies lies in their numbers. Masses of ants pour over prey, biting and slicing all at the same time. Unlucky grasshoppers or spiders are overwhelmed because they are attacked by thousands of soldiers at once. Lanchester's square law helps make sense of this outcome. Just like soldiers concentrating their fire on an opposing army, the simultaneous attack of thousands of ants can overwhelm even the best of fighters. But termite mounds eradicate this numerical advantage, since only a few ants can fit into each entrance at a time. Narrow gateways convert the battlefield from one of mass attack to one of individual confrontations, akin to dung beetles defending tunnels.[22] In this type of fight, the better-armed soldier prevails.

Army ant soldiers have big heads and crazy jaws, but the weapons of termite soldiers are even bigger. That's because army ants have to balance multiple tasks, while termite soldiers do not. Ant soldiers march long distances from their nest during an attack, racing to overtake prey. For them, selection for mobility balances selection for bigger heads and jaws, resulting in a compromise. Termite soldiers, on the other hand, hardly move at all. The one thing they do is block the gatehouse and bite everything that approaches. Termite soldiers are bigger and stronger than siafu, and, inside their rigid tunnels, they win.

Termites hold their ground, and the masses of ants move on in search of easier plunder.

Both types of city are safe as long as their walls hold. But when these barriers get broken, bad things happen. For Lachish, it took the coordinated efforts of the world's most powerful army to crash through the walls. For the termites, it took an aardvark.[23] Weighing in at 140 pounds—ten million times heavier than a termite soldier—aardvarks are walking bulldozers, lumbering beasts with strong legs and long claws specialized for digging into termite mounds. Aardvarks tear through the side of a mound, slurping up termites with their long, sticky tongues. They amble on once they've had their fill, but the holes they dig take time to repair. If an exposed mound is discovered by siafu, the result can be disastrous. As soon as the walls are breached, tactical advantage returns to attackers, since they've got the greater numbers. Invading armies swarm in, and cities perish.

• • •

Throughout this book I've compared animal weapons to human weapons.[24] I've drawn parallels between the historical processes surrounding their evolution, including the environments in which they function, the forces of selection shaping their performance, and the ways in which they change through time. In particular, I've suggested that the circumstances conducive to extreme weapons—the ingredients triggering an arms race—and the sequence of stages through which weapon evolution unfolds, are the same regardless. But how similar are these processes, really?

Teeth and horns are a part of the body of an animal. Elk antlers, for example, are manufactured as the animal develops, and the instructions relevant to their construction are written into the elk's DNA. When a bull makes sperm, the sperm carry copies of his DNA. If he succeeds in fertilizing the eggs of a cow, this DNA provides the template for building the antlers of his sons. Information encoded in the DNA is transmitted from parents to offspring and, as a result, the antlers of progeny resemble the antlers of their dads. The weapon is replicated as it is passed from father to son.[25]

Cultural traditions—instructions for everything from how to dress, how to act or communicate, and how to build shelter or weapons—also get passed from parents to offspring. And, mapped over expanses of time and place, these cultural traditions clearly change, diversifying from population to population just as body parts of animals do. But cultural information is not encoded in DNA and, for the longest time, biologists drew a line in the sand with "biological" evolution on one side, and "cultural" evolution on the other.[26] Now that line is disappearing.

Although I believe it is both illuminating and exciting to equate these processes, it's worth taking a minute to consider the differences between them. We manufacture our weapons using materials from the environment, and these structures exist as entities separate from ourselves—we can throw them away or modify them, if we want to, whereas animals are stuck with the weapons their bodies grow.

But animals manufacture structures, too.[27] Termite fortresses are perfect examples. Beaver dams, bird nests, spiderwebs, and mouse burrows are all manufactured structures. They're not parts of the bodies of the animals that make them, but they're replicated just the same. Information relevant to their construction is transmitted from one individual to the next and, as a result, the same structures are erected by generation after generation. Often this information is inherited genetically, but occasionally the details of manufacture are learned

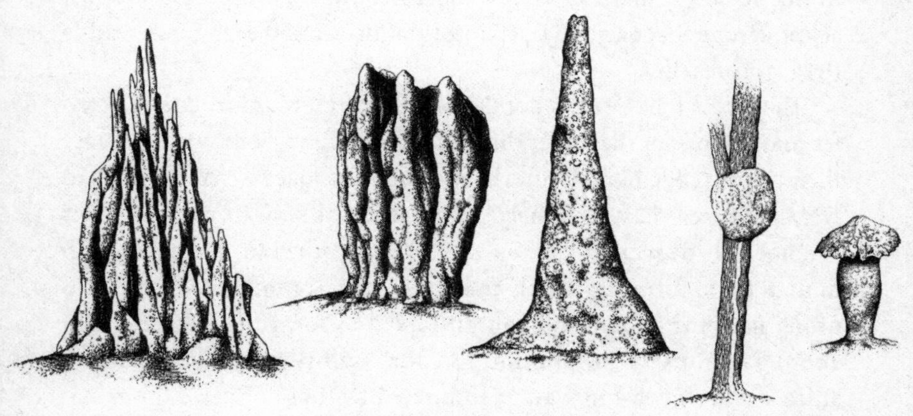

Termite fortresses

through apprenticeship and careful practice—precisely the same processes that occur in humans. All of these structures evolve.

A second difference between cultural and biological evolution is that cultural information can be transmitted more widely and rapidly than DNA. Cultural information is usually transferred from parents to progeny, but it doesn't have to be. Instructions for manufacturing an effective rifle or castle can be taught to an apprentice, carried by garrisons to foreign lands, or even stolen by spies. Since cultural information is learned, rather than inherited, it can be transferred from person to person more liberally than DNA. At least, that's what we used to think.

It turns out that transmission through DNA is far less strict than biologists initially realized. The sequencing of genomes of more and more species has revealed that fragments of DNA get swapped all the time.[28] Bacteria gobble up bits of DNA from other species—even distantly related species, such as viruses, plants, and animals—just like spies steal secrets from foreign governments or corporations. A fifth of the bacterial genome may be borrowed chunks of foreign DNA. It's easy to ignore bacteria when we consider the world around us. After all, they're really tiny and hard to tell apart. But the reality is that there may be as many as ten million bacteria species worldwide, including roughly forty thousand species thriving inside your body at this very moment.[29] The majority of living things on our planet are bacteria in one form or another, so any concept of biological evolution must acknowledge that bacteria swap information around even more readily than cultures do.

The fact is that DNA is not the only medium of information transfer, and this means that other things, too, can evolve. Some viruses house their genetic code in molecules of ribonucleic acid, or RNA, rather than DNA. Viruses most definitely evolve (they also swap parts of their genomes all over the place, as when pieces of avian flu mixed with human flu to form the deadly influenza strain of 1918, or when pieces of pig flu, bird flu, and human flu mixed to form the "swine" flu of 2009).[30] Scripts of programming code—self-replicating units in a silica world—evolve in ways stunningly like those of natural populations, despite having no RNA or DNA.[31] Consequently, although a means of information transfer and replication is essential for evolution, any number of mechanisms will suffice.

A few things do truly distinguish biological and cultural evolution, however. In biological systems the ultimate source of new variation is mutation, copy errors incorporated into the genetic code when DNA replicates itself during cell division. Mutations don't happen very often, and when they do, they strike at random.[32] Novelty in the design of manufactured weapons can arise through random events, too, such as mistakes that crop up during production. But manufactured weapons are usually modified on purpose, as engineers and designers seek to improve them. Brilliant minds like those of Archimedes, Leonardo da Vinci, and J. Robert Oppenheimer, all strived to develop new and better weapons, and the varieties they experimented with were deliberate. This means that new variation can arise faster in cultural traits than in biological ones, and this variation is more likely to be constructive. Once in place, however, variation is variation, and selection drives evolution just the same.

The most important difference is that success of a cultural trait is not linked to the reproductive success of people who use it. When elk antlers evolve, they do so because some individuals produce more offspring than others. Antlers are copied through reproduction, and winners in the battle for breeding make more copies of their antler type than losers do. Antlers evolve as the elk evolve—the processes are inextricably linked, since the mechanism of replication of the weapons is the same as the mechanism of replication of the elk. When we refer to a "population" of antlers, we also mean the population of elk that grow and wield them.

This is not true for cultural evolution. Consider the rifle. Rifles are manufactured in shops or factories, rather than wombs. Instructions for their manufacture are copied and transmitted from one person to the next, but they are recorded in documents rather than DNA. Most important of all, whether a lot of copies are made or a few has little to do with numbers of offspring. Success of one rifle type over another is not linked to the reproductive success of the people who make and use it, and the population of rifles—all rifles in existence at any point in time—is not the same as the population of people.

Cultural evolution thus unfolds on a different plane from biological evolution. Occasionally the events in these two layers intersect—the annihilation of Lachish certainly affected the reproductive success of its inhabitants, for example—but most of the time, the processes unfold in isolation. Humans evolve, and our weapons evolve, but these events unfold independently. As long as this distinction is clear, there's no reason we can't compare the evolution of animal weapons to the evolution of weapons we build.

• • •

By the end of WWI, it was clear that infantry needed a new weapon. Engineers endeavored to combine the portability of a rifle with the rapid-fire capabilities of existing machine guns.[33] The first of what would become the "assault" rifle, the Russian Fedorov Avtomat, never went into mass production because the cartridges it used, pistol ammunition, were too low-powered to be accurate beyond a few hundred feet. The French Chauchat fared better, with roughly 250,000 rifles produced by the end of the war. But this gun used ammunition that was too powerful, making the recoil uncontrollable during automatic fire. The French Ribeyrolle 1918, the Danish Weibel M/1932, and the Greek EPK soon followed, all using newly developed intermediate-caliber ammunition and thus balancing the need for accuracy at several hundred yards with the need for minimal recoil and control. But these weapons were cumbersome and heavy, as was the American M1918 Browning Automatic Rifle (BAR). By 1942 the Germans introduced the MKb 42 (H) and the Stg 44. Three years later the United States had modified the M1, adding a twenty-round detachable magazine to the rifle and giving it both selective and automatic fire options, but the ammunition for these rifles was still too powerful (the United States later scaled this down to the M16, which uses intermediate-caliber ammunition).

In 1949, the Russian Avtomat Kalashnikova, or AK-47, entered this mix of competing assault rifles. It combined the best of these other models into a frame that was virtually indestructible. The AK-47 used intermediate-caliber cartridges and had a curved, detachable magazine for smooth firing. The barrel was shorter than those of earlier assault rifles, and it was significantly lighter. Best of all, this rifle could be produced rapidly and cheaply. Simple to build, simple to use, and reliable under even the most extreme conditions, the AK-47 swept the field. Now, more than sixty years after its introduction, the AK-47 has spawned a family of related assault rifles that together are the most recognizable and abundant firearms on earth, with an estimated one hundred million units manufactured—one assault rifle for every seventy people on the planet.[34]

Even at a glance, it's apparent that the history of assault rifles has all of the elements of an evolutionary process. Although the information relevant to rifle construction is transmitted through documents and computers, rather than DNA, the result is faithful copying of the design from rifle to rifle. Rifles coming off of an AK-47 assembly line are all AK-47s and not M16s or Stg 44s. Yet, through accident or design, engineers are constantly adjusting rifle design, probing possibilities, and testing variations on the theme. Most of these experiments fail, but every now and then new design features work, and are rapidly incorporated into newer models. Most important of all, the realities of markets and battlefields act like agents of selection, culling assault rifles that are too expensive to produce, that jam or misfire, or that are more cumbersome or awkward than other available alternatives. The conditions of modern warfare shape the evolution of assault rifles in much the same way that natural agents of selection, such as battles between rival males, shape the evolution of elk antlers.

As long as we focus on the weapons—antlers and rifles—rather than the animals or humans using them, comparison of their respective evolutionary paths is relevant and informative. It's only when we start confounding rifle evolution with human evolution that things get murky. Assault rifles clearly affect the survival of humans. After all, they're used to kill people. Dead people don't have babies, so rifles also affect reproductive success. But these are not the metrics that matter for rifle evolution. What matters for the success or failure of a specific type of rifle is how well it performs relative to the other rifle models in existence at the same time. The same can be said for ships, castles, and catapults. Designs that work are copied and spread; those that fail are abandoned. It's the conditions conducive to *weapon* evolution that I compare between animals and ourselves; these conditions, I'm suggesting, are essentially the same.

• • •

My first encounter with Mayan fortifications happened by accident. In 1990, I spent two weeks in a tent in the Belize jungle, part of Princeton's tropical ecology course. It rained incessantly both weeks, and the trenches we dug around our tents couldn't divert the water fast enough. Mud plastered our faces, our clothes, our sleeping bags, and all of our equipment as we sloshed and slurped about in rubber gum boots. A tarp stretched between trees sheltered the "kitchen," and another thrown together from thatched palm fronds made up the "lab." This was my first stint in a tropical forest and, aside from almost slicing my thumb off with a machete and getting stung by a scorpion one night through the floor of my tent, the trip was a success. I was hooked. Our task for the class was to design and conduct an experiment tackling a biological topic of our choice. But on my first day exploring I stumbled into a tangled wonderland that blew my mind, and I convinced my professors to let me do something very different instead.

Hidden deep in the forest a mile from our camp was a lost Mayan city. Looming up from the dark floor of the forest were fifty-foot-tall pyramids—lots of them—all buried in detritus. Centuries of tropical growth and decay blanketed each pyramid; they sprawled like earthen hills covered with roots and vines, trees sprouting up from their sides. A thousand years earlier this had been the center of a thriving metropolis, the courtyard at the heart of an ancient city. Now, it was enshrouded in forest and forgotten.

For my project I was going to map this Mayan city. Utterly inexperienced in cartography, I set out with a compass and a sketchbook and started pacing. Each day I marched through the forest, sketching the towering mounds that rose before me, pulling aside jungle tangles, scrambling through branches, and slipping in the mud and rain. I could tell from holes dug into their sides that some of the mounds had been looted. Here and there stood a lone stone tablet—a signpost, or stela—rising from the soil in front of a pyramid, proclaiming deeds of some long-dead leader. By the end of my trip I'd found more than twenty-five pyramids nestled in that patch of Belize forest.

Author's 1990 sketch of La Milpa

The Mayan city is called La Milpa (I wasn't the first to discover it), and, two years after I wandered through its plazas, formal excavations began. It now appears that La Milpa was inhabited from 400 BCE to 850 CE, peaking with a population of roughly seventeen thousand.[35] The city perched atop a steep-sided escarpment so that, like Lachish, it sat above the surrounding plain. Aside from its location, however, La Milpa wasn't very heavily protected. Even the much larger, neighboring city of Tikal was protected only by a deep ditch and low wall. These two cities reflect the pinnacle of one of the world's most impressive ancient civilizations, a people whose society revolved around warfare. Why weren't they better defended?

The history of human fortifications is a boring one in most regions, marked by relatively little change. Most armies could not sustain the cost and logistics required for mounting effective siege attacks in far-off lands and, even when armies were large enough, the landscape sometimes got in the way. Thousand-foot precipices and spectacular gorges made approaches to Andean cities impractical, while swamps and vast tropical forests hindered movements in Mesoamerica. Wheeled towers and catapults weren't practical in these environs and, despite having the requisite wealth, forces, and political organization, the Incan, Olmec, Mayan, and Aztec empires never incorporated siege weapons into their military strategies.[36]

Without the threat of siege weapons, there was no need to enlarge defensive walls. Simple walls sufficed, and little evolution of fortification design occurred. From the earliest traces of civilization, pre-Incan villages dating to 5000 BCE, until the arrival of the Spaniards in the 1500s, fortifications in Central and South America consisted of ditches, earthen mounds, and stick or stone palisades.[37] Most cities were walled and a few, like Tenochtitlan, were surrounded by water, but basic fortifications held against the types of threats they encountered. The same holds true for much of Asia, Africa, and North America; ditches and timber palisades look almost the same in eighteenth- and nineteenth-century Iroquois and Maori villages as they did in the Fertile Crescent and the Andes more than seven thousand years ago.[38]

Only in the Middle East, Europe, and parts of Asia did siege warfare escalate, and it's here that fortresses evolved to their greatest size and complexity.[39] The protruding towers and balconies we saw in Lachish selected for a means to breach walls at a distance, and by the time of the Hellenistic Greeks artillery was added to siege repertoires.[40] Onagers, wheeled contraptions with a spring-loaded giant sling or wooden spoon, hurled rocks several hundred yards. Giant, multiperson catapults fired iron-tipped spear-sized missiles. Boulders punched holes into walls and snapped battlements. The best hits were on towers, since the corners were more fragile than the rest. Hitting a corner just right could rip away chunks of the tower, and often the walls above the gash then collapsed.[41]

The devastating impacts of artillery, in turn, selected for ways to keep these catapults out of range, beyond the reach of the flying stones. New walls were thrown up outside of the original walls. Outer walls were very expensive, since they encircled a much larger area than inner walls, and many were several miles long. Simple barriers wouldn't suffice, since they'd be smashed or undermined. So outer walls had to be equipped with full suites of defensive paraphernalia, including crenellated bastions, protruding balconies, and towers at regular intervals. For a while, these new walls worked. But the ability to hold siege weapons at greater distances only led to the design of bigger and bigger artillery. By the time of the Romans, onagers were hurling hundred-pound stones more than a thousand yards.[42]

Over time, fortress walls got thicker and taller, and siege towers and battering rams bigger. Early Assyrian siege towers had two or three stories, carrying rams handled by several dozen men. Later models got taller and wider, and, in their heyday, towers stood more than ten stories high, pushed by more than two thousand men. Battering rams reached lengths of 150 feet, crewed by a thousand men apiece.[43]

On and on the race continued as walls, towers, rams, and catapults grew. By the Middle Ages a new style of catapult was invented, the counterweight trebuchet. Trebuchets used gravity and leverage to hurl stones from a sling, and they surpassed all artillery weapons of their day. More accurate and powerful than onagers, trebuchets could launch 350-pound boulders far enough to reach the inner fortress towers of most castles. Square corners once again proved vulnerable, and they all but disappeared. Round, cylindrical towers were less prone to collapse if they were

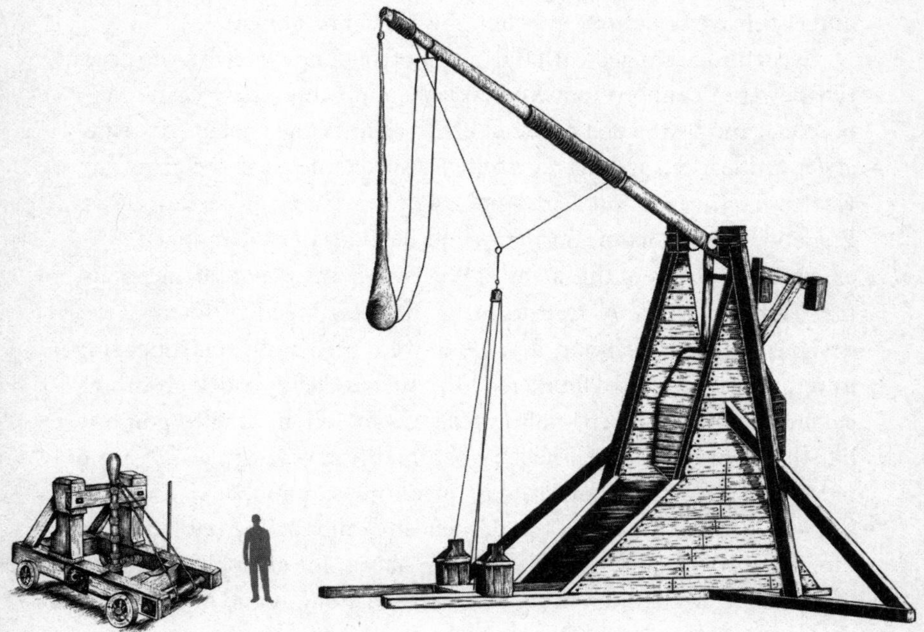

Artillery evolved in tandem with fortress size and design, hurling bigger stones farther distances. Greek onager (left) and medieval trebuchet (right).

struck directly. Even better, they were almost never hit directly, since their rounded edges deflected missile boulders to the sides.[44]

By the thirteenth century every self-respecting nobleman in the Middle East and Europe had erected a castle, and more than thirty thousand dotted the landscape.[45] Castles quickly evolved into the extravagant structures most of us are familiar with today. They had concentric outer and inner walls punctuated at regular intervals with round towers; battlements and arrow slits throughout; protruding balconies with floor holes for dropping incendiaries onto attackers; giant side-by-side gatehouses restricting access to narrow, heavily guarded, tunnel-like entryways; and massive portcullises—wrought iron gates—dropped as needed to block passage. Castles were almost always situated strategically to lie above the surrounding countryside and to be as inaccessible as possible. Often they straddled rock cliffs or were surrounded by water. If no cliffs or lakes were available, moats were built. In their day, castles ranked among the most resplendent and expensive structures ever built by the hands of man.

Everything changed with the invention of gunpowder. The destructive power of cannons rendered even the most impressive castle walls obsolete, and by the end of the fifteenth century the benefits of castle-style fortifications weren't worth their exorbitant costs. The arms race was over. The cannon had won, and thousands of castles across England, France, Spain, Germany, and Belgium were abandoned.[46]

From the ashes of this arms race grew a new style of fortifications, the "star fort." Utterly transformed, this new breed of fortress was designed to defeat cannon fire. Gone were high walls and imposing towers. These fortresses hunkered low, surrounded by shock-absorbing earthen berms, and their walls jutted outward in long, angled points—like the points of a star.[47] The logic behind the new design was to avoid presenting any broad surfaces to incoming cannonballs, since the sloping, pointed "bastions" could deflect cannonballs fired from any direction. Fort Bourtange in the Netherlands, for example, is a maze of low angled walls alternating with earthen mounds and moats. From the ground, nothing sticks up high enough to hit with a cannon; from the air, the fortress looks like a snowflake.

Star forts sprang up across Europe and in colonial settlements, including the New World. In a classic illustration of the ever-present balance between costs and benefits, British and Dutch colonies invested in earthworks and stone star forts only in locations where they were likely to encounter cannons. Forts Crown Point, Ligonier, Ontario, and Frederick, for example, all overlooked harbors or inland waterways, locations well within reach of the broadsides of warships from the major naval powers.[48] Fortifications erected inland protected settlers from natives rather than navies, and returned to the timeless tradition of inexpensive, simple wooden palisades, punctuated occasionally by protruding towers or balconies.[49]

Even star forts became obsolete with the spread of exploding artillery. Much as they spelled the end for wind-powered warships, rifled cannons with deep-penetrating exploding shells proved more than star fortresses could handle. Bombs dropped from planes soon followed, and by the time of the strategic bombing initiatives of WWII, no aboveground structure was safe from attack. Security lay deeper and deeper belowground, and the new norm became dispersed networks of tunnels and bunkers.[50]

During the Battle of Britain in 1940, the safest shelters were the deepest ones, and because they were in short supply, as many as 150,000 Londoners huddled in a labyrinth of subway tunnels each night. The Maginot Line, an inordinately expensive, thousand-mile-long string of fortifications built by the French to stall and divert the advancing Germans, was almost entirely belowground. The Japanese carved elaborate networks of tunnels, gun ports, and barracks into the volcanic rock of the Pacific Islands.[51] Shielded by solid stone, these positions withstood repeated bombardment from battleship artillery and bombs dropped by planes, and in the end most had to be cleared by brutal hand-to-hand combat. Even Gibraltar, Eisenhower's command post for the Allied invasion of north Africa, was buried deep beneath a mountain.

Once the age of aboveground fortifications ended, it didn't return. To this day, rogue organizations such as al-Qaeda and the Taliban hide in mazes of caves buried deep inside mountains,[52] and our own government maintains multibillion-dollar bunkers, such as Cheyenne Mountain, lurking hundreds of feet below the surface, shielded by millions of tons of solid rock.

• • •

Artillery and fortifications are manufactured structures, yet they evolve just like the weapons of animals do. Advances in the effectiveness of artillery select for new and better designs of fortresses, and vice versa, in a back-and-forth cycle that can spiral into an arms race. Because fortresses are fixed in place, rivals in these races have defined roles as attackers and defenders, much like animal predators and prey. The final examples pit attacker against attacker, in matched contests more reminiscent of battles between beetles or elk.

13. Ships, Planes, and States

We weren't supposed to do it, but every once in a while I'd take a canoe across Gatun Lake into the main channel of the Panama Canal. At night I could paddle right up to the container ships as they chugged past. This was the folly of youth, counting coup with sideways floating skyscrapers. The challenge was to get as close as possible. If I slid in *really* close, within, say, six feet or so, I could tap the side with my paddle and ride the big wave peeling off from the bow.

There's no way anyone on the ships ever saw me, in part because I did this under cover of darkness, but also because the people steering were more than a city block away. These were big boats. The decks alone could house three football fields end to end, if they weren't piled high with thousands of semitruck-sized boxes, and the "cabins"—hundred-foot-tall buildings ringed with glowing windows on the top floor—perched all the way back at the stern. From where the bow sliced into the water, the pilot was almost a quarter mile away.

I'd lurk at the edge of the channel, rocking gently on the swells, and wait for each ship to approach. The soft thrum of their turbines would get louder, and the black shadow of their bulk bigger, until they towered over me. White foam splashing softly in the darkness marked the exact location of the bow and, with that as a guide, I'd close to within thirty feet and let the ship come to me. Then, just as the leading edge glided past, walls rising five stories high, tractor-sized anchor poking through a hole fifty feet over my head, I'd paddle as hard and as fast as I could toward the flank. Up to the wall I'd slip, thump paddle on steel, turn, and out I'd go, straight into the five-foot wave. The ships motored on unaffected, oblivious to my silly sport, foam contrails lingering behind them as they disappeared into the distance.

Never before, or since, was the evolution of vehicle size laid more starkly before me than on those exhilarating nights in a canoe. The ships I played chicken with weighed more than two million times more than my canoe, and they displaced 130 million more pounds of water. Ship designs had come a long way, and selection for efficient transport crafted these gargantuan cargo carriers.

• • •

Vehicles are a lot like animals. They burn energy as they travel, and, like animals, they balance bulk and weight with agility and speed. Their shapes evolve over time in response to demands of particular terrains or tasks. Some become specialized for transport, others for speed, still others for fighting. When vehicles start to fight with other vehicles, selection favors characteristics that improve performance in these contests. Often, as with scramble contests or outright chases, this means faster speeds and greater agility, precluding elaborate armor or big, heavy guns. Occasionally, however, when conditions are just right, vehicles with the biggest weapons win. Bigger becomes better, and vehicles get sucked into an arms race.

Vehicular arms races require the same three ingredients as animal arms races, but it's trickier to see. In animals, the first two ingredients—competition and economic defensibility—are essential because without them, males have no reason to fight. Together, they create incentive for intense battles: the reason males invest in weapons in the first place.

Given strong incentive, the third ingredient then selects for increases in weapon size. Duels cause bigger weapons to fare better than smaller ones, pushing weapon evolution toward greater and greater extremes.

For vehicles such as ships or airplanes, the first two ingredients are provided by the states that build and use them. Governments fighting other governments justify and motivate a war, causing ship to attack ship. They load and cock the gun. But it's the details of the confrontations that determine whether the trigger gets pulled—whether bigger vehicles outperform smaller ones—and here, as with animals, the final ingredient tends to be duels.

• • •

After centuries of stasis, a single change in technology forever altered the behavior of ancient Mediterranean warships. Around 700 BCE, a simple pole of cast bronze, mounted at the waterline of the bow, turned what had been vessels of transport into weapons.[1] Galleys, powered primarily by people pulling oars, began to charge into the sides of other galleys, attempting to breach their hulls and sink them. Ships fought rival ships at close range, and one-on-one, fulfilling the final ingredient of an arms race; faster ships prevailed. Speed required lots of oars, and more oars meant bigger ships.[2]

Shipbuilders added oars, added rowers to each oar, and even added entire tiers of oars. In just a few centuries slender ships 10 feet wide and 90 feet long, powered by fifty men, jumped to double-hulled megaships 420 feet long, powered by four thousand men.[3] The biggest ships were magnificent, but they'd gotten so big their weight offset the advantage of extra oars. They were too large to be fast, and too awkward to actually close in on and ram their rivals; the pendulum of ship evolution had swung too far and these ships were nautically worthless. For a while they were given new functions. The largest ships had catamaran-style double hulls with wide, stable decks, so they were used to transport catapults and other artillery.[4] But by this point the benefits of the biggest ships were not worth their excessive costs. The arms race stalled and the pendulum swung back toward smaller ships, settling eventually on "fives." One hundred forty feet long and powered by three hundred oarsmen, fives were large and lethal, but not excessive or awkward.

Their basic design survived for more than a thousand years after this first naval arms race ended.[5]

Not until the sixteenth century did technology and one-on-one-style naval engagements realign to trigger another race. This time it was sparked by the development of a sailing warship, the galleon. Ships propelled by wind rather than oars had sturdy hulls capable of withstanding storms, and they could be handled by smaller crews, so stored provisions lasted long enough for extended voyages on the open ocean.[6] Sail-powered ships transformed exploration and commercial shipping, permitting nations with the largest navies to plant colonies around the globe.

Early galleons had a pair of cannons mounted on their bows, modern equivalents of the battering ram (the first galleons actually carried rams as well, even though these ships were ill-suited for ramming).[7] But ships were long and slender, so only a few cannons could fit at the front. Additional cannons could have been added along the upper decks, but they would have been too tippy; weight up on the deck made ships unstable. However, the invention of closeable gun ports—wooden flaps that could be shut tight to keep water out in a storm—permitted cannons to be mounted low on the sides of ships, down near the waterline where they actually made the ships more stable.[8]

Sailing galleons could now fire broadsides with lots of cannons, but they had to turn sideways to do it. Ships had to expose their flanks to opponents. They also had to fire from very close range. Smooth-bore cannons were inaccurate under the best of circumstances, but at sea, with rolling hulls shifting back and forth, even the best crews could hit only targets a few hundred yards away. Most of the time, they needed to be much closer than this to have any chance of hitting other ships. Ships started to fight rival ships at very close range—"yardarm to yardarm"—in ship-versus-ship duels and, once again, circumstances became conducive to an arms race.[9]

Bigger cannons inflicted more damage than smaller ones, and more cannons were better than fewer. More and bigger cannons required bigger ships, so the sizes of galleons grew. One row of cannons became two, and then two became three. In the fifteenth century warships packed as many as 60 cannons each, but soon they carried 74, and then

100, or even 120, and by the end of the eighteenth century some ships carried as many as 140 cannons.[10] The size, extravagance, and cost of sailing warships climbed until a new style of gun—rifle-barreled cannons with exploding shells—rendered wooden-hulled warships obsolete.[11] Like the oared galleys before them, the biggest warships were now no longer worth the price.

For both naval races, a change in ship behavior brought rivals together in new styles of interactions that favored the larger ship. Like beetles beginning to guard burrows or bull caribou starting to lock antlers head-on with rival bulls, bigger ships began to be more effective than smaller ships. Size mattered, and from that point forward it always paid to have the bigger ship. Also like animals, the ship size that performed best was relative—"big" really meant "bigger than everyone else." The playing field evolved as the ships evolved, with each advance on one side promptly matched and then bested by the other in a back-and-forth cycle that fueled the race. Eventually, the biggest ships just weren't worth the cost, and the spiral of growth collapsed. Although the details of the confrontations may vary—speed, for example, may be more important than size—the same basic dynamic of one-on-one contests helps explain all sorts of vehicular arms races, from galleons and dreadnoughts to tanks and airplanes.

• • •

A mere ten years after Orville and Wilbur Wright soared over sand dunes in Kitty Hawk, North Carolina, aircraft started shooting down other aircraft in battle. At the outset of WWI, airplanes flew reconnaissance missions, skimming over battlefields to record troop movements and locations of major artillery.[12] A biplane constructed from cloth and wood, powered by a single propeller engine, could reach speeds of one hundred miles per hour, and information obtained from pilots proved invaluable to commanders in the trenches below. The problem was that both sides appreciated the significance of aerial reconnaissance, and each attempted to prevent the other from gathering it. It didn't take long before planes from one side started encountering planes from the other, and pilots tried all sorts of clever tricks to force their opponents from the sky. Some hurled bricks at the cockpits of enemy planes; others dangled strands of rope or chain in front of the propellers.[13] Many started carrying pistols, firing at enemy pilots as they flew past.

The first attempts to mount machine guns onto airplanes fared badly; bullets whizzed straight through the spinning propeller, splintering wood when they hit the blades. The French tried a crude fix, mounting steel wedges to the inside of each blade to deflect the bullets, but it was the Germans who finally solved the problem, when they mechanically coupled propellers with the firing mechanism of the machine gun, synchronizing bullets so they shot in between the blades of the propeller.[14] Within a matter of weeks the French had copied, and then improved upon, this design, and for the remainder of the war both sides flew planes with forward-facing machine guns.

Now pilots could challenge other pilots directly, and dogfighting was born. Aerial duels pushed planes to their performance limits, and pilots quickly learned the capabilities of both their own and their opponents' machines, often attempting to exploit subtle differences in rela-

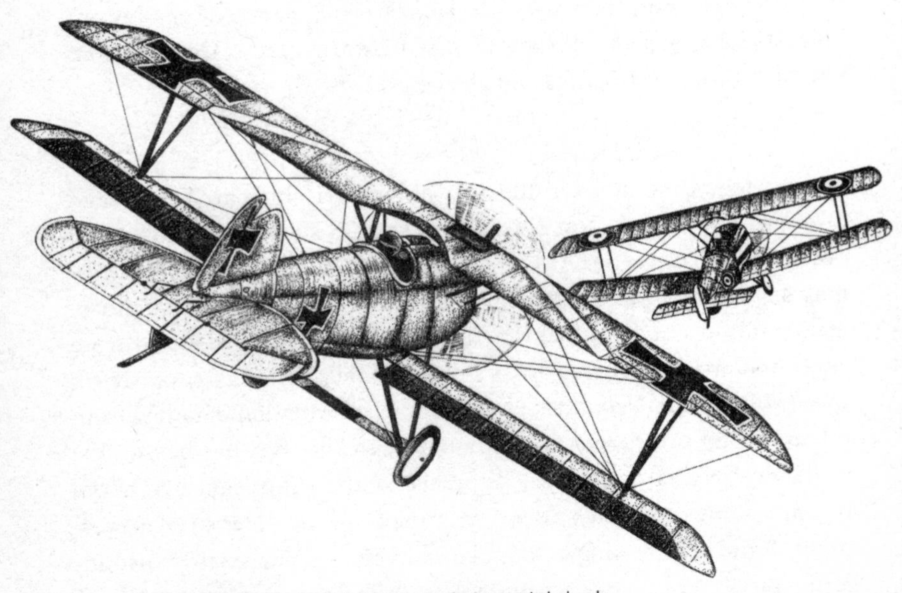

The first fighter aircraft attacked rivals in aerial duels, sparking an arms race in aircraft design.

tive speed, climb rate, or turning radius.[15] Bigger wasn't better, but faster speeds and greater maneuverability were, and planes became locked into an arms race just the same. Pilots could occasionally work around limitations of their machines, using clever tactics, skill, or trickery, but the advantage clearly went to the pilot with the superior plane, and each side raced to develop better and better machines. Back and forth the advantage went, as the Germans, then the French and British, and then the Germans again, took the mechanical lead. Model after model hit the skies, each a bit faster and better than the last.[16]

By the start of WWII, the shapes and styles of military aircraft had diversified, as planes became increasingly specialized for particular tasks. Transport planes were different from reconnaissance planes; fighters were different from lightweight fighter-bombers; and all of these were different from heavy bombers. But fighters still challenged enemy fighters in battles for control of the skies so, for these planes, the race for speed and agility surged on.

By the end of the war, propeller-driven fighter aircraft such as the U.S. P-51D Mustang could attain speeds of 440 mph, and the German Me 262, the first jet-engine fighter, could fly faster than 500 mph.[17] By the Korean War, U.S. F-86 Sabres were dropping out of near supersonic dives to fight high-speed duels with Russian-built Chinese MiG-15s.[18] Shortly thereafter, Mach II–enabling afterburner technology and air-to-air missiles further revolutionized fighter design. Now planes were moving so fast that the "G-forces" they generated during turns pushed the physical capacities of pilots to the limit—any faster and pilots blacked out, crashing multimillion-dollar aircraft in the process.[19] In modern fighters like the F-16 Fighting Falcon, pilot actions are integrated with sophisticated computer software to control planes during flight (called "fly-by-wire" systems), preventing planes from generating G-forces beyond pilot tolerances.[20]

Ironically, fighters soon became limited by how *slowly* they could fly. Supersonic they could handle, but many dogfights required close-range maneuvers that could only be accomplished at speeds under 500 mph. At these speeds planes often stalled, prompting development of "supermaneuverable" fighter aircraft such as the Russian Su-30 Flanker and the U.S. F-22 Raptor, each with rotatable nozzles permitting jet propulsion to be angled in different directions during flight.[21]

August 28, 2013, marked the one hundredth anniversary of the first documented dogfight, when the British pilot Norman Spratt, flying an unarmed Sopwith Tabloid biplane, forced a German Albatros C.I two-seater into the ground.[22] Since then, in one of the most rapid of all races, fighter aircraft have surged into supersonic, supermaneuverable beasts with stealth capabilities; sophisticated electronic flight controls, navigation, and targeting; guided air-to-air and air-to-ground missiles; and antimissile defensive measures. But this race, too, has approached its zenith.

The greatest limit to modern fighter aircraft is the pilot. The newest planes are forced to perform at levels well below what they are capable of. In fact, the primary function of computer-enhanced controls is to hold the plane back, so that the pilot doesn't pass out.[23] Unmanned aeriel vehicles (UAVs), or drones, don't have these constraints, and already they are replacing traditional aircraft for countless military tasks. UAVs cost tens of millions of dollars less per vehicle than F-16s or F-22s, and even smaller and cheaper planes—"micro air vehicles" with six-inch wingspans—are currently in the works.[24] In the not-so-distant future, piloted fighter aircraft may no longer be worth the cost.

WWII also sparked rapid evolution of bomber aircraft, incidentally, but these planes faced very different challenges. Unlike fighters, who could dodge, roll, or climb as needed to engage rival aircraft, bombers had to fly in straight lines in strict formation. As bombers neared their targets, it was imperative that they held to a constant speed and altitude, so that bombardiers—the men who actually pushed the button to drop the bombs—could aim the bombs onto targets below. So critical was this constancy that pilots handed control of the aircraft over to the bombardier during the final stages of the bombing run. That way, panicked pilots couldn't alter the course even if they wanted to.[25]

Constant flight speeds made bombers predictable targets—"sitting ducks"—not unlike fixed townships and cities, and in many ways the evolution of these planes paralleled the evolution of walled fortifications and castles.[26] Bombers did not attack rival bombers, like fighters attacked fighters. Instead, their survival depended upon defense, thwarting the advances of enemy fighters. In these confrontations, exposed planes were dead planes, so rotatable, protruding turrets with machine guns were added to the top and bottom, and guns were added to the nose, tail, and flanks. As with castles, the idea was to leave no flank undefended, and bombers soon carried guns and crew sufficient to provide covering fire to all sides of the craft. Even the names of these planes reflected the defensive logic of their design, such as the B-17 "flying fortress," or the B-29 "superfortress." Unfortunately for the crews, these planes still had to remain light enough to fly, so armor plating was limited and the effort to build fortresses of the air short-lived.

In contrast with fighters, bombers flew straight paths over targets, rendering them especially vulnerable to enemy fire. The evolution of bombers resembled that of castles, as planes became increasingly fortified with defensive turrets and machine cannons positioned to cover all angles of approach.

• • •

The grandest scale at which arms races can unfold is between rival states. States are even more like animals than vehicles, gobbling up resources and competing with each other for control of these resources. In the ancient world, the most limiting resources, besides people, arable land, water, and living space, were deposits of copper and tin.[27] Metals were hard to come by, and nearly all went into weapons. Today we still fight over arable land, water, and living space; but we also fight over access to energy, primarily in the form of oil. States depend on natural resources for their survival. There aren't enough to go around, and the result, not surprisingly, is competition. But *how* states compete—who challenges whom, whether confrontations turn into arms races, and whether arms races escalate to outright war—is more predictable than most people realize.

When states compete with states, the "individuals" that confront each other are rival governments, and the relevant weapons their respective military forces. New states are born from time to time, and other states disappear, but this isn't the evolutionary turnover that matters here, because arms races between states happen much faster than this, beginning and ending during the political lifetimes of the involved states. Rather, it's the military establishments within each state that grow or recede. Circumstances conspire to spark an arms race when rivals face each other in such a way that bigger militaries suddenly become much better than smaller ones. The state with more or better weapons gains an advantage that prompts the other side to catch up, launching them both into increasingly extravagant back-and-forth cycles of military spending. States continue to spend, shunting more and more of their resources into weapons, and militaries expand, until the race culminates in outright war; or until one of the states spends beyond what it can sustainably afford and fiscally collapses.

For political arms races, the best parallel with animals is not the gradual turnover in populations we've focused on thus far, but rather a confrontation between two rival males—two crabs, for example, that face off in the sand. Who will back down? An arms race between states erupts in the same way a contest between crabs escalates; neither yields, pushes give way to grabs and jabs, grappling to smashing, and then pounding, and finally unrestricted war. When it comes to warfare, states behave just like crabs on a beach.

Pick any period in history and glance at a political map. You'll see big states and little states, and everything in between. Some states are born rich—their borders include vast natural resources, favorable climate, and secure, defensible terrain. Others have almost nothing. Wealthy states have larger total resource pools, or gross domestic products (GDPs), than poorer states, and they can afford to spend more on weapons. To put this in perspective, in 2011 the United States had a GDP of roughly $15 trillion.[28] At that time, this was twice the GDP of China, eight times the GDP of Russia, thirty times the GDP of Iran, and *four hundred thousand* times larger than the GDP of Montserrat and Tuvalu. States differ spectacularly in how much they have available to spend.

Like crabs, states have mandatory expenses that they must pay first, before they can shunt resources into weapons. Crabs are made from millions of cells, which need to be fed and protected. If the cells die, the crab dies, and most of its mandatory expenditures revolve around keeping these cells alive. States are made up of people, and necessary expenses protect and nurture these people—things such as education and welfare, police forces and highways. Only when funds are left over after mandatory expenses are accounted for can states invest in militaries, weapons, and other luxury items.[29]

The richest few states have vast sums of discretionary resources that they can pour into weapons development, technological development, ships, planes, munitions, training, and personnel. Most states have much smaller discretionary pools. Many have no discretionary funds at all, and any spending they devote to militaries erodes dangerously the funds they need to survive.[30] States invest in militaries to the extent that they are able, but the sizes of militaries vary widely from country to country. As with beetle horns, caribou antlers, and fiddler crab claws, the relative size of a state's military provides an honest signal of its fighting strength: a perfect tool for deterrence.

But the parallels with animals run even deeper. States posture just like crabs, waving their weapons and advertising military strength for all to see. They challenge each other incessantly, pushing and shoving, rubbing claw to claw, probing for weakness. And, as with crabs, most encounters end before they really begin. When one side is stronger than the other, the weaker state either stands down or is overtaken; either way, the conflict ends before it has a chance to escalate.

Huge militaries are effective deterrents, and a tiny state would no more attack a superpower outright than a tiny crab would escalate a fight with a leviathan. Instead, tiny states challenge other tiny states. Midsized states confront midsized states, and only big states face off with big states. Despite a political landscape littered with countries of all sizes, contests tend to coalesce around clusters of equally matched opponents. Occasionally, when a confrontation settles into a true duel with matched rivals squaring off face-to-face, and when both sides have discretionary resource pools large enough to sustain it, a state-versus-state contest can spiral into an arms race.

• • •

It was apparent the moment the United States dropped atomic bombs on Hiroshima and Nagasaki that a game-changing new weapon had arrived. The Hiroshima bomb contained 140 pounds of uranium 235, exploding with the destructive equivalent of 16,000 tons of TNT (16 kilotons). In a single flash of light an entire city vanished, along with 150,000 people.[31]

After the war, countries scrambled to develop nuclear weapons, but for most the technology and cost were prohibitive. As the costs climbed higher, fewer and fewer countries could stay in the game, and a complex post-WWII political landscape began to coalesce around just two opposing superpowers. As the Warsaw Pact and NATO each pulled member states into their respective folds, the world stage aligned into a one-on-one showdown between the United States and the Soviet Union, perfect preconditions for an arms race. For almost forty years both powers poured their discretionary budgets into weapons development, building incomprehensibly excessive arsenals.

Weapon after weapon got swept into the race, as the Soviet Union and the United States each ratcheted up technology on all fronts simultaneously. Early in the Cold War the Soviets amassed a tremendous fleet of submarines, more than 450 boats. Their fleet outnumbered ours several times over, but in 1955 we upped the ante by launching the first nuclear-powered submarine. The USS *Nautilus* was for all intents and purposes invisible. It could stay submerged for months at a time, crossing oceans without ever breaking the surface. In one step we'd rendered their massive fleet obsolete, and the Soviets began developing a whole new fleet of nuclear submarines of their own.[32]

U.S. fighter planes fared poorly in Korea when they first faced off against the faster and more maneuverable Soviet-made MiG-15s,[33] so the United States rushed to develop new airplanes, a technological surge that led to a string of supersonic fighters—the "century series"—including the F-100 and the F-106. Our planes, in turn, prompted the Soviets to develop their own supersonic fighters, resulting in the MiG-21, the MiG-23, and the Su-15.[34]

Back and forth each side went. Tanks on both sides got bigger and stronger—thicker armor, faster engines, and much bigger guns, but in the 1960s the United States added a much better vehicle to the mix. The Sheridan tank could be dropped by parachute from airplanes, and it could work as an amphibious vehicle in water; best of all, it could fire a guided antitank missile from its main gun instead of artillery shells.[35] Although the Sheridan never did perform as promised, it was a game-changing advance in technology, and the Soviets promptly embarked on a research program to design a tank like this of their own. Soviet tank developments surged, as the T-34 turned into the T-54, the T-55, and then the T-62, -64, -72, and the T-80. They built tanks with antitank missiles, tanks with antiaircraft missiles, and all sorts of antimine and bridging vehicles. By 1980, the Soviets had amassed an armada of more than 120,000 armored vehicles.[36]

Naval fleets mushroomed in size; bomber fleets expanded in capability and numbers; fighters grew faster and deadlier, and each side built more and more of them. But the fastest evolving weapons by far were the nuclear warheads and the vehicles needed to deliver them. Early on, nuclear warheads had to be dropped from the air by bombers, but by the 1950s both sides were experimenting with placing warheads into the nose cones of missiles, and the idea of intercontinental ballistic missiles (ICBMs) was born.[37]

In 1957 the Soviets launched a satellite into space, triggering the "space race" between the two powers. On the surface, the space race was civil and nonmilitary. Beneath the surface, it was yet another facet to the arms race. With Sputnik the Soviets proved beyond any doubt that they possessed missile technologies capable of delivering nuclear warheads anywhere in the world, catapulting the United States into full-speed missile development.[38] Rocket technologies whirled forward, as each side experimented with propulsion, fuels, and guidance systems. At the same time, advances in the warheads made them both deadlier and smaller—all the better to pack into the tops of rockets.

By the 1960s both countries were already reeling from the escalating cost, but they surged forward nevertheless, and the numbers of warheads and missiles climbed.[39] Emphasis shifted from first strikes to retaliatory strikes, as both states considered how best to defend themselves against a first strike from the other. The most obvious solution was more missiles. The best way to be sure that some missiles survive is to have more missiles than the other side.[40] Missiles massed in one place were too easy to hit, so both sides scattered their arsenals across the landscape in dispersed belowground silos.

Even better, keep your missiles moving. The Soviets mounted launchers onto railcars so they could be shuttled from place to place, and both countries began putting missiles on submarines. Submarines all but guaranteed a retaliatory strike capability, since they were always moving and invisible, and therefore impossible to target. Bombers flew rotating shifts, so that some were airborne all the time, and by the late 1970s the United States was developing "stealth" bombers that would be just as invisible as submarines. Railcars, hidden silos, submarines, and bombers all combined to form a swirling maze of missile platforms, making them very difficult to hit.[41]

New and better rockets replaced older rockets; single warhead missiles were replaced by multiple independent reentry vehicles (MIRVs), with three or more separate, independently targetable warheads per rocket; guidance systems began to incorporate terrain-matching posi-

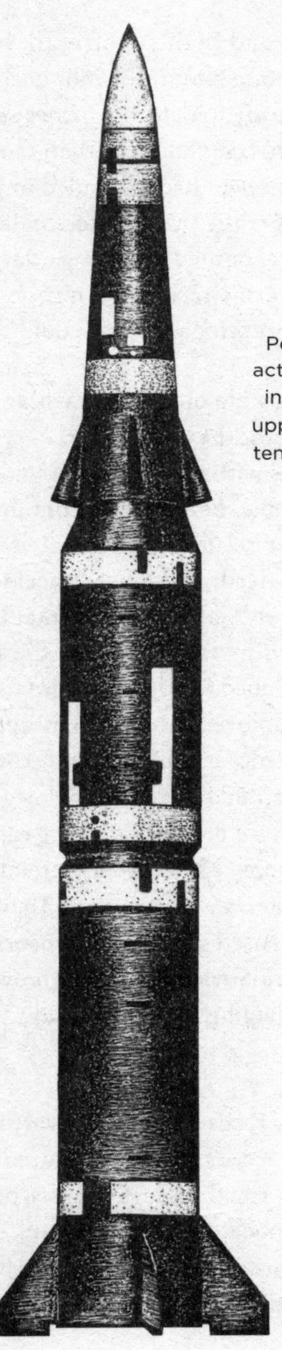

Pershing II missiles ran on solid fuel with active radar guidance. Deployed to Europe in the mid-1980s, these "bunker busters" upped the ante in the Cold War, heightening tensions between the United States and the Soviet Union to critical levels.

tioning systems, improving accuracy; and both countries developed extensive electronics and radar detection systems for monitoring the missiles of the other side, poised and ready to detect launches seconds after they took place. By the 1980s each side had more than ten thousand warheads, and the warheads themselves had expanded in killing power from kilotons to megatons—all while fitting into smaller and smaller packages, enabling them to be mounted on a larger variety of delivery platforms.[42] Weapons of this arms race could now exterminate civilization on a planetary scale; in principle, they could do this many thousands of times over.

In 1983, the United States upped the ante once again, with a solid-fueled nuclear missile that had state-of-the-art guidance systems, allowing it to match radar video signals with internal maps to achieve "pinpoint" accuracy.[43] Light and portable, Pershing IIs contained an adjustable-yield warhead that could be set to detonate between five and fifty kilotons, depending on the need, and their placement in Europe—much closer to the Soviet Union than the continental United States—meant drastically reduced warning times. Finally, the Pershing II was earth penetrating. It was designed to be a "bunker buster."

The Soviets saw this new missile as an overt attempt to incapacitate their command and control structure, since it could be on top of them in as little as six minutes, and it penetrated hardened belowground bunkers. But the Soviets, meanwhile, were building a new weapon of their own—an automated launch system that could fire retaliatory missiles even after the command center was destroyed. The "dead hand," as it was called, was a satellite-based network of sensors that, once activated, could automatically fire Soviet ICBMs to previously targeted coordinates without a human having to push a button.[44]

• • •

The Cold War trumped all other arms races in history. Every major weapons system got sucked into this race, as each side poured more and more of its budget into military growth, and the costs of each weapon soared. A submarine in the 1960s cost $110 million; by the 1980s it cost $1.5 billion—more than a ten-fold increase in price. Bombers went from $8 million to $250 million over the same period.[45] All of the weapons exploded in cost simultaneously—guidance systems, missiles, fighters, tanks, cruisers, aircraft carriers, warheads—the list goes on. The cost of Cold War arsenals climbed so high that a typical year's military spending for either superpower exceeded the entire GDP of dozens and dozens of countries.[46]

Only the United States and the Soviet Union could afford to play this game, and during that time nuclear weapons were the ultimate deterrent. Deterrence during the Cold War worked just like it does in animals, with each side using early low-risk stages of confrontation to size each other up. Political scuffles on the world stage coalesced into polarized conflicts between the two superpowers, one side backed by NATO and the other by the Warsaw Pact. Like crabs pushing against the claw of a rival, conflagrations in Korea, Vietnam, Afghanistan, and the Middle East—the so-called proxy wars—provided low-risk conventional means for the superpowers to flex their muscles.[47] Aggression by one side was matched by pushback from the other, and each conflict dissipated before escalating into a full-fledged nuclear war. Although it's no consolation to the hundreds of thousands who lost family members or friends in these proxy wars, compared with the planetary annihilation that could have been unleashed had either side pushed their buttons, deterrence during the Cold War actually brought relative peace.

Many authors have tried to estimate the economic cost of the Cold War arms race, and it's proven to be a difficult thing to measure.[48] All agree the price was staggering. The United States, for example, spent trillions of dollars on defense during this period, as much as 10 percent of its GDP each year, and up to 70 percent of its discretionary budget.[49] Funds allocated to the military had to come from somewhere, and inevitably these monies were diverted away from other uses; social welfare programs, education, health care, and housing all suffered during the Cold War as a direct result of military spending.[50] The Soviets paid even more, easily 15–17 percent of their GDP and, by some estimates, as much as 40 percent—devastating levels of military spending.[51] In order to stay in the race, the Soviets were exceeding their discretionary resource pools. Like Irish elk leaching calcium and phosphorus from their skeletons to pay for massive antlers, the Soviets were dipping into the resources they needed to stay alive. So severely were social programs depleted that nearly every aspect of Soviet society suffered.[52] This pattern of spending was not sustainable, and in December 1991, the Soviet Union collapsed. Latvia, Estonia, Belarus, and Ukraine declared independence from Russia; the Union of Soviet Socialist Republics officially dissolved; and Gorbachev resigned as president, declaring the office extinct. The most dangerous arms race in the history of the world ended without a nuclear war.

It very nearly didn't.

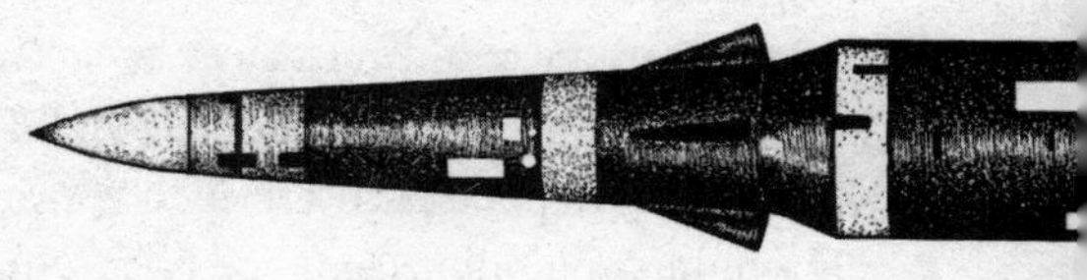

14. Mass Destruction

November 8, 1983. Panic shot through the Soviet high command. Tensions between the United States and the Soviet Union had been climbing for weeks, and the Soviets were now on the highest state of alert since the 1962 Cuban missile crisis. In May, they had activated intelligence operatives throughout Europe who monitored the daily activities of critical NATO personnel so that they could pick up subtle changes in behavior that might indicate an impending attack. Now, reports were flooding in from all directions at once. NATO bases had sprung to full alert, and the joint chiefs and heads of state were being sequestered in war rooms. Intercepted NATO communications had switched to unfamiliar formats, possibly disguising hidden messages, and now they'd all gone "radio silent." More alarming still, it appeared NATO forces might be at DEFCON 1—their most dire state of nuclear alert. The worst fears of the Soviets were materializing; nuclear first strike was imminent.[1]

As terrifying as this was, it wasn't unexpected. For the past year the

United States had steadily dialed up the intensity of its actions, sneaking submarines in close to Soviet shorelines, listening and testing just how close they could get. Aircraft carriers launched fighters that soared straight at Soviet airspace, tripping all the sensors and sending local bases into frenzied panic, only to veer away at the last possible moment. It would soon be possible for NATO forces to launch Pershing II missiles from Europe; the first were supposed to arrive later that year, but a few might have been smuggled in early. Even *one* of these bunker busters would be enough to incapacitate Soviet leadership, so every confrontation from this point forward had to be considered a possible first strike.

Again and again the United States charged, sometimes with fighters, other times with bombers, always choosing flight paths designed to take them straight into Soviet airspace, always flying close enough to trip alarms and place the military on high alert, and always turning away at the last possible second.

On April 6, six U.S. Navy aircraft crossed the line, buzzing over Zeleny Island, one of the Kuril Islands controlled by the Soviets. The Soviets were furious and, to save face, they promptly retaliated with fighter flights over the Aleutian Islands. Each false alarm fed the tension within the Soviet high command, and base commanders got jumpier and jumpier. Nerves were frayed to the breaking point.

On September 1 an aircraft flew straight into Soviet airspace without turning back, crossing directly over an ICBM testing range on the Kamchatka Peninsula, and on toward their Pacific Fleet headquarters at Vladivostok. The Soviets, already on edge, shot it down—a bad decision. Korean Air Lines Flight 007 plummeted into a frigid sea, killing all 269 passengers and enraging the world, pouring gasoline onto an already out-of-control political fire.

Then, on September 26, a ballistic missile alarm flashed. The Soviets' state-of-the-art early warning detection system signaled a Minuteman ICBM launch from within the continental United States. All systems at the base went into instant overdrive, but Lieutenant Colonel Stanislav Petrov, on call during that shift, made the executive decision to stand down. The ground radars had not confirmed the launch, and it made no sense that we would fire only a single missile (it was later determined to be a computer error).

One month later, six thousand U.S. combat troops, with fourteen thousand more held in support, deployed to Grenada. With the whole world watching, the United States ousted Soviet-supported forces in just nine days, a stinging loss of political face reminiscent of the Cuban missile crisis. Given the "anticommunist" rhetoric surrounding the invasion, the Soviets feared it might be the first in a string of imminent NATO attacks on Warsaw Pact holdings. Nicaragua or Cuba could be next.

Then, on November 4, two days after Grenada, Soviet attack submarine *K-324* tangled in a sonar array cable towed behind the U.S. Navy frigate USS *McCloy*. The sub was forced to surface off the South Carolina coast and be towed, in full view of the U.S. Navy, for the next four days toward Cuba—another shocking humiliation for Soviet leaders. More important, *K-324* had been tracking activities of several U.S. ballistic missile submarines. Because of this mishap, the U.S. subs had vanished. *Four days* passed, and still they were missing. With that much time they could be anywhere—within strike distance of Mother Russia, for example. Now, November 8, intelligence reports were pouring in: NATO bases on alert, heads of state sequestered in war rooms, military communications gone dark. Was this it? The moment everyone dreaded—was it happening?

Soviet forces sprang to full readiness, and air units in Poland and East Germany mobilized for immediate takeoff. Some thought it was the beginning of an attack; others thought it was just a military practice operation; still others thought it was a military exercise designed deliberately as a trick—a cover for an actual first strike. Soviet plans for a first strike incorporated just such a ruse, a military "exercise" used to masquerade an actual attack, so why should U.S. strategy be any different? The problem was, the "dead hand" was not yet operational, so the only way the Soviets could guarantee a retaliatory counterstrike was to launch their missiles *first*—a preemptive launch—but if they were to do this there wasn't a moment to spare; they had to launch now. Should they? No NATO missiles had yet been fired; what if it wasn't an actual attack? To make matters worse, the Soviets were at that moment facing a dangerous leadership vacuum—General Secretary Yuri Andropov was on his deathbed, and for the past two months he had not attended Politburo meetings. Now, in this stunning moment of crisis, political rivalries and leadership uncertainty added to the chaos. The fate of the world hung in the balance as the Soviet command debated whether to push the button.

• • •

On that November day when life on our planet nearly ended I was in high school in Ithaca, New York, focused on an experiment for my biology class that wasn't working and memorizing a handful of lines for my role in that year's school play; I'd been cast as the nerdy valedictorian in the musical *Grease*. I was oblivious to the imminence of my near annihilation. We all were.

My future father-in-law had just retired as the director of operations at Davis-Monthan Air Force Base, home of Titan II nuclear intercontinental ballistic missiles. You'd think that he, at least, would have been aware. He was not. In fact, it wasn't until two years after the fact, when KGB agent Oleg Gordievsky defected and related the full story to British intelligence, that anybody outside of the Soviet Union learned just how close we'd all come to nuclear war. Now, with the declassification of Cold War documents, the full story has started to emerge,[2] and the lesson is terrifying indeed. For a second time—the first being the Cuban missile crisis—humanity teetered on the brink.

It's probably not an exaggeration to say we all owe our lives to a handful of people in a Soviet war room. Somehow, the leaders managed to make the right decision that day. Despite a string of close calls, repeated provocation, political turmoil from the absence of their leader, and inaccurate information, they did not launch their missiles, opting instead to hold their breath and watch as the deadliest waiting game in the history of the planet unfolded.

As suddenly as the buildup started, it ended. Leaders came out of their bunkers, and bases resumed normal activity. NATO hadn't *actually* been at DEFCON 1. They'd been conducting a military exercise—a simulated state of nuclear attack—called Operation Able Archer 83.

The United States had launched a massive, coordinated, military "dry run" of a full nuclear first strike.[3] They just hadn't bothered to tell the Soviets, and nobody realized they nearly precipitated Armageddon. The most horrifying fact of the Able Archer 83 crisis was that NATO wasn't even aware it existed.

• • •

I'm no expert in geopolitics or national security. This should be self-evident, since I work on beetles. But I've come this far comparing arms races of animals and people, so it seems only prudent to reflect for a moment on the Cold War and its legacy. What can animal weapons teach us about the world we live in today?

Many feel we survived the Cold War because of deterrence, and after researching this book I'm inclined to agree. We escaped two extremely dangerous near misses, but in the end deterrence prevailed. The threat of total destruction prevented either superpower from launching, and it kept all the other states at bay in the process.

It's been more than two decades since the Cold War ended. During that time, the United States has reigned as the sole superpower. The United States has larger arsenals, a bigger navy, and better air forces than any other state, and we spend billions of dollars yearly to keep it that way. We're the biggest crab left on the beach, and we have fantastic weapons. Are we safer as a result?

In some ways, we probably are. Modern conventional weapons appear to behave just like animal weapons. State-of-the-art supersonic, supermaneuverable fighters, brand-new *Gerald R. Ford*–class supercarriers (due to arrive in 2015), and unprecedented satellite surveillance systems and intelligence-gathering supercomputers all cost a fortune to design, build, and maintain. These are weapons technologies only the richest states can afford, and they confer a pronounced advantage in battle. As with the largest, most expensive animal weapons, our conventional forces no doubt deter rival states from initiating unrestricted war against us.

Just as the unchallenged supremacy of the Roman army and the British navy each resulted in episodes of relative peace (the Pax Romana and Pax Britannica, respectively), so some have argued that U.S. military dominance has ushered in a "Pax Americana."[4] Until another state can match us militarily and economically, the risk of an outright conventional attack on our forces is small. With full-scale war off the table, the only options left to our rivals are asymmetric—sneaky tactics that break the rules.

States that could never attack us outright snipe incessantly at our soldiers and our will to fight with guerrilla tactics. Suicide bombs, car bombs, and IEDs undermine the effectiveness of our conventional weapons. These attacks are irritating, and deadly to a few, but none directly threaten the sovereignty of the United States or the security of the vast majority of its inhabitants. On the surface, then, deterrence appears to be working just fine, and comparisons with animal weapons suggest the extreme cost of our modern arsenals may be justifiable.

The problem is weapons of mass destruction.

• • •

The sad fact is the Cold War arms race left us with weapons unlike any that have ever existed, placing us in uncharted and dangerous territory. Modern nuclear and biological weapons are unfathomably destructive. It's tough to imagine the sudden death of billions of people. What would that postapocalyptic planet look like? Climate, crops, forests, food—all would be irrevocably altered in ways catastrophic to humankind. Biodiversity: gutted; ecosystems: shattered; basically, every aspect of life as we know it, including all the people we know and care for, as well as all the others we've never heard of, would be ashes and dust.

This sounds like the stuff of Hollywood, but it's not fiction.[5] Collateral damage from weapons of mass destruction is likely to be staggering, and this changes the stakes of conflict. Weapons of mass destruction are so deadly that any use of them at all could threaten our very existence. Like it or not, this places us in an age where our only option is deterrence—anything else would be suicide. But deterrence has its limits, and we may be facing those limits now.

In crabs, beetles, flies, and caribou—indeed, in *all* animals with extreme weapons—deterrence works for very good reasons, and it works only when specific conditions are met. It boils down to choosing battles wisely. It never pays to shy from a fight you might win, but it often pays to walk away from the ones you're likely to lose. The trick lies in predicting the outcome beforehand. In order to do this, potential combatants must have a reliable method for evaluating each other's fighting ability.

When honest signals are present, males with smaller weapons usually walk away. Power discrepancies are obvious because weapon size tracks the health, body size, nutritional reserves, and status of each male—all the factors that matter for predicting the outcome of battle. For these animals, waving weapons works.

When honest signals are lacking, on the other hand, there is no safe way to predict the winner beforehand. Males who walk away now may be skipping battles they could win, sacrificing critical opportunities to mate. In animals, at least, when honest signals are missing, contests become sudden and dangerous, and very often deadly. If we're to take our lessons from crabs and caribou, then the prerequisite for peace is weapons that function as honest signals of fighting ability. Are they?

For animal weapons to be honest signals, they must be expensive—exorbitantly expensive. So expensive, in fact, that only the top-condition males can afford them. Costs keep signals honest. If anybody could afford big weapons, then all males would have them, and differences in weapon size would be meaningless. Only when most males cannot possibly afford them will big weapons provide reliable signals of fighting strength. Then, and only then, will it pay for males with small weapons to walk away.

Early in the Cold War weapons of mass destruction met this requirement. They were prohibitively expensive, and only the richest two superpowers had nukes. But as the race progressed, warheads got cheaper. The cost of *conventional* weapons—submarines, fighters, and carriers, for example—soared, but the nuclear warheads themselves got smaller and cheaper. Pretty soon, England and France were testing nuclear warheads, then China and South Africa. By the 1970s India had successfully tested nuclear warheads, too, and by the 1990s so had Pakistan. Now, Israel and North Korea have them as well. The most important precondition for deterrence is disappearing.

Biological weapons are even cheaper. Research was well under way during WWII, and for decades both the United States and the Soviet Union aggressively pursued ways to weaponize deadly pathogens.[6] Before the 1972 Biological and Toxin Weapons Convention banned biological weapons research, the United States was spending $300 million per year developing deadly antihuman, antilivestock, and anticrop pathogens, even experimenting with using insects to deliver the pathogens to targets.[7] Even after the ban, the Soviets strove onward, perfecting and stockpiling hundreds of tons of heat- and cold-resistant extra-deadly strains of anthrax, plague, tularemia, botulism, smallpox, and Marburg virus.[8]

Biological weapons weren't very expensive to begin with, and in recent years their price has plummeted. Today, it's possible to assemble perfect copies of the world's most dangerous diseases—pathogens such as the 1918 strain of avian influenza, responsible for roughly one hundred million deaths—in a simple basement laboratory for a few thousand dollars. If anybody can make these weapons, then anybody can use them, and this throws the essential logic of deterrence out the window.

• • •

We are racing toward a world where lots of states possess weapons of mass destruction, regardless of the size and relative strength of their conventional fighting forces. Nuclear and biological weapons break the rules—they cheat—by providing states with few resources a means to bring down wealthier rivals. If history is any lesson here, then weapons of mass destruction are likely to erode the cost-effectiveness of expensive conventional forces. Just as longbows and muskets foretold the end of medieval armor, and exploding artillery the end for sailing warships and castles, so, too, may we be nearing the point where low-cost nuclear and biological weapons spell the end for expensive conventional military forces.

But the bigger problem, as I see it, is the weapons themselves, and the collateral damage they stand to inflict. Even if deterrence *does* still work—which I'm not convinced it should—it doesn't mean these weapons won't ever be used; deterrence just means their use will be less likely. It may only be 1 out of 100 encounters or, as in caribou, 6 out of

11,600, but if animals teach us anything, it's that conflicts eventually do escalate all the way to full-fledged battle. Before the Cold War this detail wouldn't have mattered very much; now, because of weapons of mass destruction, it makes all the difference. The silver lining, I suppose, is that the conflicts most likely to spiral all the way to unrestricted war are predictable. They involve one-on-one duels between evenly matched rivals.

The Cold War arms race revolved around just such a rivalry—the two biggest crabs on the beach. But it doesn't have to be the big crabs. There are lots and lots of crabs on every beach, waving and pushing, sizing each other up, and any of these confrontations has the potential to escalate. A crab facing a much larger foe may opt to walk away, but he doesn't just disappear into the sunset. He seeks out a better battle—one he is more likely to win. This time, he'll approach a more evenly matched foe. When a midsized crab squares off against another midsized crab, early gestures fail to settle the dispute up front; push comes to shove, and both sides hold their ground; shoves intensify, turning into whacks and squeezes; finally, if neither backs down, the next step is inevitable: unrestricted war.

On a beach, if two midsized crabs launch into full battle it's not likely to matter much for everybody else. But we're not on that beach. Weapons of mass destruction are cheap enough now that most midsized states have them, or if they don't yet, they soon will, and the destructive power of these weapons is so great that their use anywhere, even once, has the potential to unravel civilization on a global scale. If weapons of mass destruction are part of the equation, then we can't let *any* confrontation escalate. Ever. But stopping them all is a tall order. A mere glance at our present political map shows hot spots with deadly potential—rivalries poised and ready to flare into full war. North Korea versus South Korea; India versus Pakistan; Israel versus Iran—all of these nations stand toe-to-toe with equally matched rivals, and all either already have or will soon have weapons of mass destruction.

• • •

Where does this leave us? For fourteen chapters I've argued that our weapons and animal weapons are similar. But this is true only up to a point. The deadliest weapons of today have no precedent—never before has an animal wielded the capacity to destroy life on such a planetary scale, and never have there been weapons so dangerous they can never be used.

Today's world is a quagmire of rivalries and factions, ethnic disputes, and religious wars, and the last thing we need is for large and unknowable numbers of these players to be armed with weapons of mass destruction. Even when the safety of the world lay in the hands of just two governments—each a superpower acutely aware of the destructive consequences of their deadliest weapons, and each having layers of safeguards in place to prevent a misfire—we still almost escalated to nuclear war in at least two instances. Now, decisions of catastrophic life and death lie in the hands of many more governments and, in some cases, even in the hands of rogue individuals. Will all of them make the right decision every time?

The picture gets even scarier when terrorist organizations are thrown into the mix. Thus far, terrorists have been armed only with conventional weapons. They may not fight "by the rules," but the weapons they wield are still pulled from standard arsenals, and the damage they're able to inflict relatively minor. What happens when one of these organizations gets their hands on a weapon of mass destruction?

Writing this book has pulled me a long way from rainforests and beetles, mud, rain, and elk. I started this venture with stories to tell of the most magnificent animals in the world. Along the way I ventured deeper and deeper into human history, enthralled, and at times appalled, by what I learned about our past. I stand awed and shaken—thrilled by the parallels and, at the same time, terrified by the prospects. For me, the final message is clear. Weapons of mass destruction change the stakes, and the logic, of battle. We're not likely to survive another arms race.

Acknowledgments

A project of this scope would not have been possible without support from colleagues, family, and friends. First and foremost, I thank my family: my wife, Kerry, and my children, Cory and Nicole, for bearing with me as I became ever more consumed by this process. I thank my colleagues, collaborators, and students, who picked up slack while I hid in the stacks of the library writing, especially Cerisse Allen, Laura Corley-Lavine, Annika Duke, Ian Dworkin, Hiroki Gotoh, Erin McCullough, Devin O'Brien, Jema Rushe, Jennifer Smith, Ian Warren, and Robbie Zinna. I thank the National Science Foundation, especially Zoe Eppley, Irwin Forseth, Dianna Padilla, Adam Summers, Kimberlyn Williams, and William Zamer, for funding my research program. The NSF plays an absolutely vital role in supporting basic research. They are the lifeblood of scholarship in the United States, and none of my work would have been possible without them.

Writing a book of this nature required a great deal of "deprogramming." I had to strip away all the stylistic rules I teach my students so assiduously. Twenty years of scripting grant proposals and academic papers for technical journals proved to be a hindrance, rather than an asset. I had to start over—learn to write anew. This was a refreshing (and exhausting) experience, and I could not possibly have managed it without the patience and critical feedback of my editor, Gillian Blake, her assistant, Caroline Zancan, and my colleague and friend Carl Zimmer, all of whom took me to task on multiple occasions.

This work benefitted greatly from feedback from many additional people. For critiquing all or parts of this manuscript in its various forms, I thank Brett Addis, Harrison Ambrose III, Harrison Ambrose IV, Katharine Ambrose, Tina Bennett, Alexis Billings, Kelly Bright, Kerry Bright, Ray Bright, Kristen Crandell, Annika Duke, Cory Emlen, Natalia Demong Emlen, Stephen Emlen, Daphne Fairbairn, Harry Greene, Melissa Hamre, Matthew Herron, Erin Kuiper, Tara Maginnis, Christine Miller, Devin O'Brien, Alison Perkins, Mike Ryan, David Tuss, and Carl Zimmer. Katharine Ambrose, Kerry Bright, Stephen Emlen, Erin McCullough, Alison Perkins, and David Tuss, in particular, worked through the entire manuscript more than once, providing feedback that dramatically improved the final text.

David Tuss was a pleasure to work with, patiently bouncing illustrations back and forth with me, reworking and, at times, completely redrawing figures until we both were thrilled with the outcome. For sharing details of their research experiences, I thank John Christy, Gerald Wilkinson, and David Zeh; for guiding me into the world of publishing, I thank Ben Roberts; and for first suggesting that I delve into the literature on human weapons, I thank Alison Kalett.

Military history turned out to be a vast and overwhelming literature, and a few authors stood out for me. The writings of Robert O'Connell, in particular, were transformative as I stepped, tentatively at first, and then with ever greater tenacity, out of the world of biology and into that of military history. In many ways, O'Connell has done in his books the same thing that I attempt here, but in reverse. His expertise is in military history, yet he reaches into biology as needed to provide perspective. I write about biology, dipping into military history for precisely the same reason. For readers interested in this interface of biology and history, I highly recommend his books *Of Arms and Men:*

A History of War, Weapons, and Aggression (Oxford: Oxford University Press, 1989) and *Soul of the Sword: An Illustrated History of Weapons and Warfare from Prehistory to the Present* (New York: Free Press, 2002). I also thank him for generously taking the time to read through this manuscript, correcting many of the details of my coverage of military history, particularly material related to the Cold War.

I also highly recommend Trevor Dupuy's book *The Evolution of Weapons and Warfare* (New York: Da Capo Press, 1984), which helped put the major transitions in military technology into perspective for me; John Keegan's *The Face of Battle: A Study of Agincourt, Waterloo, and the Somme* (London: Penguin Books, 1983), which brought the realities of ancient battle to life with startling vividness and relevance; and David Hoffman's *The Dead Hand: The Untold Story of the Cold War Arms Race and Its Dangerous Legacy* (New York: Doubleday, 2009)—brilliant, poignant, and terrifying in its portrayal of the aftermath of history's deadliest arms race.

I thank *Caffé Dolce* and its wonderful staff for the perfect place to think and write, as well as for the best coffee in Missoula. Molly Bloom, Michelle Daniel, and Meryl Levavi, my Henry Holt production team, were thorough and patient—exactly what I needed. Finally, I'd like to thank my agent, Tina Bennett, and her assistant, Svetlana Katz, for unflinching support and guidance through all of the stages of this process. I could not have done this without them.

Notes

1. Camouflage and Armor

1. Oliver Pearson and Anita Pearson, "Owl Predation in Pennsylvania, with Notes on the Small Mammals of Delaware County," *Journal of Mammology* 28 (1947): 137–47; Charles Kirkpatrick and Clinton Conway, "The Winter Foods of Some Indiana Owls," *American Midland Naturalist* 38 (1947): 755–66.
2. Ibid.
3. F. B. Sumner, "An Analysis of Geographic Variation in Mice of the *Peromyscus polionotus* Group from Florida and Alabama," *Journal of Mammology* 7 (1926): 149–84; Sumner, "The Analysis of a Concrete Case of Intergradation Between Two Subspecies," *Proceedings of the National Academy of Sciences of the U.S.A.* 15 (1929): 110–20; Sumner, "The Analysis of a Concrete Case of Intergradation Between Two Subspecies. II. Additional Data and Interpretations," *Proceedings of the National Academy of Sciences of the U.S.A.* 15 (1929): 481–93; Sumner, "Genetic and Distributional Studies of Three Subspecies of *Peromyscus*," *Journal of Genetics* 23 (1930): 275–376.
4. Lynne Mullen and Hopi Hoekstra, "Natural Selection Along an Environmental Gradient: A Classic Cline in Mouse Pigmentation," *Evolution* 62 (2008): 1555–70.
5. F. B. Sumner and J. J. Karol, "Notes on the Burrowing Habits of *Peromyscus polionotus*," *Journal of Mammology* 10 (1929): 213–15; Jesse Weber and Hopi Hoekstra, "The Evolution of Burrowing Behavior in Deer Mice (genus *Peromyscus*)," *Animal Behavior* 77 (2009): 603–9.
6. Donald W. Kaufman, "Adaptive Coloration in *Peromyscus polionotus*: Experimental Selection by Owls," *Journal of Mammology* 55 (1974): 271–83.
7. Hopi Hoekstra and her colleagues have revealed how mutated versions of two genes arose and spread in the beach populations, causing animals in these areas to develop with white

fur. Fur color in mice and other mammals is controlled through regulation of the synthesis of pigments. One of the molecules involved in this process is the *melanocortin-1 receptor* (*Mc1r*), a roughly corkscrew-shaped protein that loops like a sea serpent in and out of the surface membrane of pigment-producing cells. Because folds of this protein extend all the way through the membrane, *Mc1r* can couple events happening on the outside of the cell with processes occurring on the inside.

Mc1r acts like a switch because it flips between two alternative contortional states, and this change in shape alters the production of pigment. When *Mc1r* is twisted into one of its shapes it is relatively inactive, and pigment cells produce a pale yellow pigment called "phaeomelanin." When it switches to its alternate and more active shape, then the cells begin producing a different pigment called melanin. Melanin is a deep, dark brown, and fur infused with melanin results in brown mice.

Whether *Mc1r* triggers production of phaeomelanin or melanin depends, in part, on the presence of other molecules outside of the cell. For example, binding of an activator protein to one of its outer loops can switch *Mc1r* so that cells begin producing melanin, whereas binding of an inhibitory protein can block melanin synthesis, reverting cells to production of the paler phaeomelanin. Pigment cells are riddled with thousands of copies of *Mc1r* throughout their outer membranes, all triggering the production of either light or dark pigments inside the cell. The result is a blend of the two pigments, and a continuous range of possible fur darkness. When concentrations of the activator molecules are high, most copies of *Mc1r* are active and melanin synthesis predominates. When concentrations of inhibitors are high, melanin levels plummet. In this basic fashion, shifts in the levels of activators and inhibitors cause the color of fur to vary from hair to hair, and also from mouse to mouse.

In interior populations of the oldfield mouse, as in related mouse species, the majority of copies of *Mc1r* are active during pup development, and fur on the backs of these mice is brown. But when Hoekstra's team looked at mice from the Gulf coast, they found that *Mc1r* was much less active. When coastal pups developed, less melanin was produced and fur on their backs was mostly white. Hoekstra's team compared, base pair for base pair, the DNA sequences of the genes coding for *Mc1r* and its activator and inhibitor (the inhibitor is called *Agouti*) in inland and beach mice, and they spotted two differences. Animals sampled from coastal populations carried copies of *Mc1r* and *Agouti* genes that were just a little bit different from the copies carried by inland mice. At some point in the past, mutations altered the sequences of these two genes, and these changes caused beach mice to develop with lighter fur. Hopi E. Hoekstra, Rachel J. Hirschmann, Richard A. Bundey, Paul A. Insel, and Janet P. Crossland, "A Single Amino Acid Mutation Contributes to Adaptive Beach Mouse Color Pattern," *Science* 313 (2006): 101–4; Cynthia C. Steiner, Jesse N. Weber, and Hopi E. Hoekstra, "Adaptive Variation in Beach Mice Produced by Two Interacting Pigmentation Genes," *Public Library of Science (PLoS) Biology* 5 (2007): e219; Cynthia C. Steiner, Holger Römpler, Linda M. Boettger, Torsten Schönenberg, and Hopi E. Hoekstra, "The Genetic Basis of Phenotypic Convergence in Beach Mice: Similar Pigment Patterns but Different Genes," *Molecular Biology and Evolution* 26 (2008): 35–45.

The beach mouse mutation to the *Mc1r* gene made this receptor spend less of its time in its active state, tipping the delicate balance in favor of lighter fur. The beach mouse mutation to the *Agouti* gene increased its activity, resulting in higher concentrations of circulating *Agouti* protein. Because *Agouti* inhibits the activity of *Mc1r*, mice with the beach allele of the *Agouti* gene showed lower levels of *Mc1r* activity, and pups with this mutation also developed with lighter fur. Both of these mutations lightened the color of mouse fur, and together they yielded very white mice.

8. Task Force Devil Combined Arms Assessment Team (Devil CAAT), "The Modern Warrior's Combat Load, Dismounted Operations in Afghanistan, April–May 2003," U.S. Army Center for Army Lessons Learned (2013).
9. A. Dugas, K. J. Zupkofska, A. DiChiara, and F. M. Kramer, "Universal Camouflage for the Future Warrior," U.S. Army Research, Development, and Engineering Command, Natick

Soldier Center, Natick, MA 01760 (2004); K. Rock, L.L. Lesher, C. Stewardson, K. Isherwood, and L. Hepfinger, "Photosimulation Camouflage Detection Test," U.S. Army Natick Soldier Research, Development and Engineering Center, Natick, MA (2009), NATICK /TR-09/021L.

10. Ibid.
11. Eric Coulson, "New Army Uniform Doesn't Measure Up," Military.com, April 5, 2007; Matthew Cox, "UCP Fares Poorly in Army Camo Test," *Army Times*, September 15, 2009.
12. U.S. Government Accountability Office, "Warfighter Support: DOD Should Improve Development of Camouflage Uniforms and Enhance Collaboration Among the Services," Report to Congressional Requesters, September 2012.
13. Ibid. See also L. Hepfinger, C. Stewardson, K. Rock, L. L. Lesher, F. M. Kramer, S. McIntosh, J. Patterson, K. Isherwood, G. Rogers, and H. Nguyen, "Soldier Camouflage for Operation Enduring Freedom (OEF): Pattern-in-Picture (PIP) Technique for Expedient Human-in-the-Loop Camouflage Assessment," report presented at the 27th Army Science Conference, JW Marriott Grande Lakes, Orlando, FL, November 29–December 2, 2010; Joseph Venezia and Adam Peloquin, "Using a Constructive Simulation to Select a Camouflage Pattern for Use in OEF," Proceedings of the 2011 Military Modeling and Simulation Symposium, Society for Computer Simulation International (2011).
14. A. Bartczak, K. Fortuniak, E. Maklewska, E. Obersztyn, M. Olejnik, and G. Redlich, "Camouflage as the Additional Form of Protection During Special Operations," *Techniczne Wyroby Włókiennicze* 17 (2009): 15–22; M. A. Hogervorst, A. Toet, and P. Jacobs, "Design and Evaluation of (urban) Camouflage," *Proc. SPIE 7662*, Infrared Imaging Systems: Design, Analysis, Modeling, and Testing XXI, 766205 (April 22, 2010).
15. This book focuses on morphological weapons of animals, and does not have space to cover the rich repertoire of chemical weapons. For interested readers I recommend Thomas Eisner, *For Love of Insects* (Cambridge, MA: Belknap Press of Harvard University Press, 2005), and Thomas Eisner, Maria Eisner, and Melody Siegler, *Secret Weapons: Defenses of Insects, Spiders, and Other Many-Legged Creatures* (Cambridge, MA: Belknap Press of Harvard University Press, 2007).
16. P. F. Colosimo, C. L. Peichel, K. Nereng, B. K. Blackman, M. D. Shapiro, and D. Schluter, "The Genetic Architecture of Parallel Armor Plate Reduction in Threespine Sticklebacks," *PloS Biology* 2 (2004): E109; M. D. Shapiro, M. E. Marks, C. L. Peichel, B. K. Blackman, K. S. Nereng, B. Jónsson, D. Schluter, and D. M. Kingsley, "Genetic and Developmental Basis of Evolutionary Pelvic Reduction in Threespine Sticklebacks," *Nature* 428 (2004): 717–23.
17. T. E. Reimchen, "Injuries on Stickleback from Attacks by a Toothed Predator (*Oncorhynchus*) and Implications for the Evolution of Lateral Plates," *Evolution* 46 (1992): 1224–30.
18. Michael Bell, Matthew P. Travis, and D. Max Blouw, "Inferring Natural Selection in a Fossil Threespine Stickleback," *Paleobiology* 32 (2006): 562–77.
19. Pamela F. Colosimo, Kim E. Hosemann, Sarita Balabhadra, Guadalupe Villarreal Jr., Mark Dickson, Jane Grimwood, Jeremy Schmutz, Richard M. Myers, Dolph Schluter, and David Kingsley, "Widespread Parallel Evolution in Sticklebacks by Repeated Fixation of *Ectodysplasin* Alleles," *Science* 307 (2005): 1928–33; Rowan D. H. Barrett, Sean M. Rogers, and Dolph Schluter, "Natural Selection on a Major Armor Gene in Threespine Stickleback," *Science* 322 (2008): 255–57.
20. Jun Kitano, Daniel I. Bolnick, David A. Beauchamp, Michael Mazur, Seiichi Mori, Takanori Nakano, and Catherine Peichel, "Reverse Evolution of Armor Plates in the Threespine Stickleback," *Current Biology* 18 (2008): 768–74.
21. F. Wilkinson, "Arms and Armor," *Journal of the Royal Society of Arts* 117 (1969): 361–64; Trevor N. Dupuy, *The Evolution of Weapons and Warfare* (New York: Da Capo Press, 1984).
22. Ibid.
23. Ibid.
24. F. Kottenkamp, *History of Chivalry and Ancient Armour* (London: Willis and Sotheran, 1857); Wilkinson, "Arms and Armor," 361–64; Dupuy, *Evolution of Weapons and Warfare*;

Robert L. O'Connell, *Of Arms and Men: A History of War, Weapons, and Aggression* (Oxford: Oxford University Press, 1989).

25. Wilkinson, "Arms and Armor," 361–64; Dupuy, *Evolution of Weapons and Warfare*; Dave Grossman and Loren W. Christensen, *The Evolution of Weaponry: A Brief Look at Man's Ingenious Methods of Overcoming His Physical Limitations to Kill* (Seattle: Amazon Publishing, 2012).
26. Dupuy, *Evolution of Weapons and Warfare.*
27. John Keegan, *The Face of Battle: A Study of Agincourt, Waterloo, and the Somme* (London: Penguin Books, 1983); Dupuy, *Evolution of Weapons and Warfare.*
28. Dupuy, *Evolution of Weapons and Warfare*; Grossman and Christensen, *Evolution of Weaponry.*
29. Dupuy, *Evolution of Weapons and Warfare.*
30. Grossman and Christensen, *Evolution of Weaponry.*

2. Teeth and Claws

1. S. B. Williams, R. C. Payne, and A. M. Wilson, "Functional Specialization of the Pelvic Limb of the Hare (*Lepus europaeus*)," *Journal of Anatomy* 210 (2007): 472–90.
2. Benjamin T. Maletzke, Gary M. Koehler, Robert B. Wielgus, Keith B. Aubry, and Marc A. Evans, "Habitat Conditions Associated with Lynx Hunting Behavior During Winter in Northern Washington," *Journal of Wildlife Management* 72 (2007): 1473–78; John R. Squires and Leonard F. Ruggiero, "Winter Prey Selection of Canada Lynx in Northwestern Montana," *Journal of Wildlife Management* 71 (2007): 310–15.
3. Christopher J. Brand, Lloyd B. Keith, and Charles A. Fischer, "Lynx Responses to Changing Snowshoe Hare Densities in Central Alberta," *Journal of Wildlife Management* 40 (1976): 416–28; Kim G. Poole, "Characteristics of an Unharvested Lynx Population During a Snowshoe Hare Decline," *Journal of Wildlife Management* 58 (1994): 608–18; Brian G. Slough and Garth Mowat, "Lynx Population Dynamics in an Untrapped Refugium," *Journal of Wildlife Management* 60 (1996): 946–61.
4. Ronald E. Heinrich and Kenneth D. Rose, "Postcranial Morphology and Locomotor Behaviour of Two Early Eocene Miacoid Carnivorans, *Vulpavus* and *Didymictis*," *Palaeontology* 40 (1997): 279–305; Blaire Van Valkenburgh, "*Déjà vu*: The Evolution of Feeding Morphologies in the Carnivora," *Integrative and Comparative Biology* 47 (2007): 147–63.
5. L. D. Martin, "Fossil History of the Terrestrial Carnivora," in *Carnivore Behavior, Ecology, and Evolution*, ed. J. L. Gittleman (Ithaca, NY: Cornell University Press, 1989), 335–54; Van Valkenburgh, "*Déjà vu*," 147–63; Julie Meachen-Samuels and Blaire Van Valkenburgh, "Craniodental Indicators of Prey Size Preference in the Felidae," *Biological Journal of the Linnean Society* 96 (2009): 784–99.
6. Van Valkenburgh, "*Déjà vu*," 147–63.
7. Blaire Van Valkenburgh, "Skeletal Indicators of Locomotor Behavior in Living and Extinct Carnivores," *Journal of Vertebrate Paleontology* 7 (1987): 162–82; Van Valkenburgh, "*Déjà vu*," 147–63; Julie Meachen-Samuels and Blaire Van Valkenburgh, "Forelimb Indicators of Prey-Size Preference in the Felidae," *Journal of Morphology* 270 (2009): 729–44.
8. Van Valkenburgh, "*Déjà vu*," 147–63.
9. Ibid.
10. I experienced this proclivity firsthand when I was in high school, camping with my dad in Kenya's Samburu National Reserve. We'd pitched our tent beneath a sprawling tree in a clearing by the river. The wardens stopped by while we were setting up to tell us that three weeks prior, two women had been killed in that spot by a hippo walking through their tent on the way to the water. But there was nowhere else to put the tent (that little clearing was the official campground), so there it sat.

    We left camp for an hour at sunset to drive through the park, and when we returned we found chaos. A troop of baboons had raided our tent, shredding everything in sight. They had torn through the side leaving a great, gaping hole, and they had bitten clean through two of the poles. They had even scent-marked our sleeping bags with urine. But by then it was dark and we had no place to go, so we patched the pieces together as best we could and tried to sleep.

Falling asleep in a situation like that is not easy. Modern tents rely on long, bowed poles to support them, and when these are snapped the design truly fails. The patchwork of foul-smelling fabric and string we'd salvaged was not very convincing. Also, I can attest from experience that any sense of security you derive from a tent comes from the visual barrier of fabric. Animals on the outside can both hear you and smell you, tucked away in your nylon shell. The only thing they can't do is *see* you. A predator has no way to know that the shell you've erected is less than a millimeter thick, and for this reason tents really do protect you from most things that go bump in the night (unless, of course, that thing is so huge, like an elephant or hippo, that it squashes you by accident). This night, however, our barrier was broken. We could see out and they could see in. We did finally sleep, after watching giraffe silhouettes drift by, black against the starlight in our newly torn "window."

Sometime in the night we woke to a scream, followed immediately by shrieks and hoots and loud crashing branches in the tree above us. The baboons had returned to roost fifteen feet over our heads, and a leopard was attacking them! We learned later that this happens often. Leopards like to pull baboons away from their troop while they're perched helplessly in trees at night, and there we sat directly beneath with a four-foot gaping hole in our tent. I panicked, grabbing my sleeping bag and racing for the car, which, in hindsight, was the worst thing I could have done since I was completely exposed to the leopard above my head as I ran.

11. Sharon B. Emerson and Leonard Radinsky, "Functional Analysis of Sabertooth Cranial Morphology," *Paleobiology* 6 (1980): 295–312; Martin, "Fossil History of the Terrestrial Carnivora," 335–54; Van Valkenburgh, "*Déjà vu*," 147–63; Graham J. Slater and Blaire Van Valkenburgh, "Long in the Tooth: Evolution of Sabertooth Cat Cranial Shape," *Paleobiology* 34 (2008): 403–19.
12. Van Valkenburgh, "Skeletal Indicators of Locomotor Behavior," 162–82; Blaire Van Valkenburgh and Fritz Hertel, "Tough Times at La Brea: Tooth Breakage in Large Carnivores of the Late Pleistocene," *Science* 261 (1993): 456–59.
13. Ibid.
14. Martin, "Fossil History of the Terrestrial Carnivora," 335–54.
15. P. W. Freeman and C. A. Lemen, "The Trade-Off Between Tooth Strength and Tooth Penetration: Predicting Optimal Shape of Canine Teeth," *Journal of Zoology* 273 (2007): 273–80.
16. Blaire Van Valkenburgh and Ralph E. Molnar, "Dinosaurian and Mammalian Predators Compared," *Paleobiology* 28 (2002): 527–43.
17. Van Valkenburgh and Hertel, "Tough Times at La Brea," 456–59; Van Valkenburgh, "Feeding Behavior in Free-Ranging, Large African Carnivores," *Journal of Mammology* 77 (1996): 240–54; Van Valkenburgh, "Costs of Carnivory: Tooth Fracture in Pleistocene and Recent Carnivores," *Biological Journal of the Linnean Society* 96 (2009): 68–81.
18. Francis Juanes, Jeffrey A. Buckel, and Frederick S. Scharf, "Feeding Ecology of Piscivorous Fishes," chapter 12 in *Handbook of Fish Biology and Fisheries*, vol. 1, *Fish Biology*, ed. Paul J. B. Hart and John D. Reynolds (Malden, MA: Blackwell Publishing, 2002), 267–83.
19. P. W. Webb, "The Swimming Energetics of Trout. I. Thrust and Power Output at Cruising Speeds," *Journal of Experimental Biology* 55 (1971): 489–20; P. W. Webb, "Fast-Start Performance and Body Form in Seven Species of Teleost Fish," *Journal of Experimental Biology* 74 (1978): 211–26; Patrice Boily and Pierre Magnan, "Relationship Between Individual Variation in Morphological Characters and Swimming Costs in Brook Charr (*Salvelinus fontinalis*) and Yellow Perch (*Perca flavescens*)," *Journal of Experimental Biology* 205 (2002): 1031–36.
20. Bent Christensen, "Predator Foraging Capabilities and Prey Antipredator Behaviours: Pre- Versus Postcapture Constraints," *Oikos* 76 (1996): 368–80; Frederick S. Scharf, Francis Juanes, and Rodney A. Rountree, "Predator Size-Prey Relationships of Marine Fish

Predators: Interspecific Variation and Effects of Ontogeny and Body Size on Trophic-Niche Breadth," *Marine Ecology Progress Series* 208 (2000): 229–48.

21. Susan S. Hughes, "Getting to the Point: Evolutionary Change in Prehistoric Weaponry," *Journal of Archaeological Method and Theory* 5 (1998): 345–408.
22. Michael J. O'Brien, John Darwent, and R. Lee Lyman, "Cladistics Is Useful for Reconstructing Archaeological Phylogenies: Palaeoindian Points from the Southeastern United States," *Journal of Archaeological Science* 28 (1991): 1115–36; Briggs Buchanan and Mark Collard, "Investigating the Peopling of North America Through Cladistics Analyses of Early Paleoindian Projectile Points," *Journal of Anthropological Archaeology* 26 (2007): 366–93; R. Lee Lyman, Todd L. VanPool, and Michael J. O'Brien, "The Diversity of North American Projectile-Point Types Before and After the Bow and Arrow," *Journal of Anthropological Archaeology* 28 (2009): 1–13.
23. George C. Frison, "North American High Plains Paleo-Indian Hunting Strategies and Weaponry Assemblages," in *From Kostenki to Clovis: Upper Paleolithic–Paleo-Indian Adaptations*, ed. O. Soffer and N. D. Praslov (New York: Plenum Press, 1993), 237–49; Hughes, "Getting to the Point," 345–408; Briggs Buchanan, Mark Collard, Marcus J. Hamilton, and Michael J. O'Brien, "Points and Prey: A Quantitative Test of the Hypothesis That Prey Size Influences Early Paleoindian Projectile Point Form," *Journal of Archaeological Science* 38 (2011): 852–64.
24. Hughes, "Getting to the Point," 345–408.
25. G. H. Odell and F. Cowan, "Experiments with Spears and Arrows on Animal Targets," *Journal of Field Archaeology* 13 (1986): 195–212; George C. Frison, "Experimental Use of Clovis Weaponry and Tools in African Elephants," *American Antiquity* 54 (1989): 766–84; J. Cheshier and R. L. Kelly, "Projectile Point Shape and Durability: The Effect of Thickness: Length," *American Antiquity* 71 (2006): 353–63; M. L. Sisk and J. J. Shea, "Experimental Use and Quantitative Performance Analysis of Triangular Flakes (Levallois Points) Used as Arrowheads," *Journal of Archaeological Science* 36 (2009): 2039–47.
26. D. C. Waldorf, *The Art of Flint Knapping* (Cassville, MO: Litho, 1979); Hughes, "Getting to the Point," 345–408.
27. Stuart J. Feidel, *Prehistory of the Americas* (Cambridge, MA: Cambridge University Press, 1992).
28. Buchanan et al., "Points and Prey," 852–64.
29. Ibid.
30. Lyman, VanPool, and O'Brien, "The Diversity of North American Projectile-Point Types," 1–13; Douglas H. MacDonald, *Montana Before History: 11,000 Years of Hunter-Gatherers in the Rockies and Plains* (Missoula, MT: Mountain Press Publishing Company, 2012).
31. Hughes, "Getting to the Point," 345–408.
32. Ibid.
33. Ibid.
34. Ibid.

3. Claspers, Graspers, and Giant Jaws

1. Unfortunately, our trip was short-lived. On the third day I broke out in hives, itching all over. I'd never had an allergic reaction before, but I knew what they were and I couldn't imagine a worse place to go into anaphylaxis. I travel prepared for catastrophes—I had antibiotics, painkillers, sutures, burn bandages, and a snakebite kit with me on that trip—but this was an emergency I hadn't anticipated, and I didn't have an EpiPen or antihistamines. Had my reaction been instant, as they so often are, I never would have lived. But the symptoms came on gradually. My eyelids and fingers began to swell, then my throat started to get tight. Within a few hours my voice was down to a crackly whisper and I was having trouble breathing from the swelling. That was the tipping point. We collapsed camp in a dash and, as the sun set, began racing back upriver toward Coca.

   Clever and Selfo came through for me that night. I'm sure they were terrified I would die on their watch, sinking their fledgling outfitter business in the process, but they risked their boat and their lives getting me to medicine. The Napo in those days had a "shoot-on-sight" curfew after dark—part of the war on drugs. Thick mist over the water helped hide our boat, but navigating around sandbars and snags without lights is dicey. I was delirious most of the trip, though I remember looking up at one point as a huge floating tree whooshed by in the darkness.

2. L. G. Nico and D. C. Taphorn, "Food Habits of Piranhas in the Low Llanos of Venezuela," *Biotropica* 20 (1988): 311–21; V. L. de Almeida, N. S. Hahn, and C. S. Agostinho, "Stomach Content of Juvenile and Adult Piranhas (*Serrasalmus marginatus*) in the Paraná Floodplains, Brazil," *Studies on Neotropical Fauna and Environment* 33 (1998): 100–105.
3. J. H. Mol, "Attacks on Humans by the Piranha *Serrasalmus rhombeus* in Suriname," *Studies on Neotropical Fauna and Environment* 41 (2006): 189–195.
4. F. Juanes, J. A. Buckel, and F. S. Scharf, "Feeding Ecology of Piscivorous Fishes," in *Handbook of Fish Biology and Fisheries*, ed. P. J. B. Hart and J. D. Reynolds (Malden, MA: Blackwell Publishing, 2002); J. R. Grubich, A. N. Rice, and M. W. Westneat, "Functional Morphology of Bite Mechanics in the Great Barracuda (*Sphyraena barracuda*)," *Zoology* 111 (2008): 16–29.
5. H. B. Owre and F. M. Bayer, "The Deep-Sea Gulper *Eurypharynx pelecanoides* Vaillant 1882 (Order Lyomeri) from the Hispaniola Basin," *Bulletin of Marine Science* 20 (1970): 186–92; J. G. Nielsen, E. Bertelsen, and A. Jespersen, "The Biology of *Eurypharynx pelecanoides* (Pisces, Eurypharyngidae)," *Acta Zoologica* 70 (1989): 187–97.
6. Gavin J. Svenson and Michael F. Whiting, "Phylogeny of Mantodea Based on Molecular Data: Evolution of a Charismatic Predator," *Systematic Entomology* 29 (2004): 359–70.
7. H. Maldonado, L. Levin, and J. C. Barros Pita, "Hit Distance and the Predatory Strike of the Praying Mantis," *Zeitschrift Für Vergleichende Physiologie* 56 (1967): 237–57; Taku Iwasaki, "Predatory Behavior of the Praying Mantis, *Tenodera aridifolia* II. Combined Effect of Prey Size and Predator Size in the Prey Recognition," *Journal of Ethology* 9 (1991): 77–81; R. G. Loxton and I. Nicholls, "The Functional Morphology of the Praying Mantis Forelimb (Dictyoptera: Mantodea)," *Zoological Journal of the Linnean Society* 66 (2008): 185–203.
8. Sheila N. Patek, W. L. Korff, and Roy L. Caldwell, "Deadly Strike Mechanism of a Mantis Shrimp," *Nature* 428 (2004): 819–20.
9. D. Lohse, B. Schmitz, and M. Versluis, "Snapping Shrimp Make Flashing Bubbles," *Nature* 413 (2001): 477–78.
10. Beautiful studies of the genetics of caste development reveal how chemical cues interact with developmental hormones to regulate the expression of genes in ways that are specific to each caste, coordinating the details of their growth. For example, Ehab Abouheif and Greg A. Wray, "Evolution of the Genetic Network Underlying Wing Polyphenism in Ants," *Science* 297 (2002): 249–52; Julia H. Bowsher, Gregory A. Wray, and Ehab Abouheif, "Growth and Patterning Are Evolutionarily Dissociated in the Vestigial Wing Discs of Workers of the Red Imported Fire Ant, *Solenopsis invicta*," *Journal of Experimental Zoology, Part B: Molecular and Developmental Evolution* 308 (2007): 769–76. There are also a number of excellent studies of the effects of nutrition and hormones on caste development, such as Diana E. Wheeler, "The Developmental Basis of Worker Caste Polymorphism in Ants," *American Naturalist* (1991): 1218–38.
11. Sheila N. Patek, J. E. Baio, B. L. Fisher, and A. V. Suarez, "Multifunctionality and Mechanical Origins: Ballistic Jaw Propulsion in Trap-Jaw Ants," *Proceedings of the National Academy of Sciences* 103 (2006): 12787–92.
12. Olivia I. Scholtz, Norman Macleod, and Paul Eggleton, "Termite Soldier Defence Strategies: A Reassessment of Prestwich's Classification and an Examination of the Evolution of Defence Morphology Using Extended Eigenshape Analyses of Head Morphology," *Zoological Journal of the Linnean Society* 153 (2008): 631–50.

13. Trevor N. Dupuy, *The Evolution of Weapons and Warfare* (New York: Da Capo Press, 1984); R. L. O'Connell, *Of Arms and Men: A History of War, Weapons, and Aggression* (Oxford: Oxford University Press, 1989); M. van Creveld, *Technology and War: From 2000 B.C. to the Present* (New York: Free Press, 1989); O'Connell, *Soul of the Sword: An Illustrated History of Weaponry and Warfare from Prehistory to the Present* (New York: Free Press, 2002).
14. Robert L. O'Connell, *Sacred Vessels: The Cult of the Battleship and the Rise of the U.S. Navy* (Oxford: Oxford University Press, 1991); Robert Jackson, *Sea Warfare: From World War I to the Present* (San Diego: Thunder Bay Press, 2008).

4. Competition

1. I'd put the bond between father and son to a pretty good test the week before. While my dad was in town I dragged the canoe from its hiding place in the grass by the tree and canoed to the territories alone, keeping up the jacana vigil. Crosswinds were stiff that morning, and I'd had to kneel low in the middle of the boat, paddling as hard as I could to fight the gale and keep the boat on course. As I tucked in beside one of the territories I came into the lee of tall trees on the far shore, and the wind died. I dropped anchor, assembled the tripod and scope, and set to work, bare feet tucked beneath the bench seat, knees splayed to stabilize my weight in the bottom of the boat. Black flies had been everywhere that week, and it didn't take long before I felt the telltale tickle of one crawling up my calf. I'm still not sure why I bothered to look, since flies can be swatted without taking your eye from the scope. But I did look, and I flinched. Crawling up my bare leg was a tarantula, not a fly. Six inches of spider! I admit it; I reacted badly. I jumped, the canoe rocked, and a thousand dollars' worth of tripod and scope flipped over the side, vanishing forever into the swift muddy deep.
2. C. Yeung, M. Anapolski, M. Depenbusch, M. Zitzmann, and T. Cooper, "Human Sperm Volume Regulation: Response to Physiological Changes in Osmolality, Channel Blockers, and Potential Sperm Osmolytes," *Human Reproduction* 18 (2003): 1029.
3. J. Rutkowska and M. Cichon, "Egg Size, Offspring Sex, and Hatching Asynchrony in Zebra Finches, *Taeniopygia guttata*," *Journal of Avian Biology* 36 (2005): 12–17.
4. W. A. Calder, C. R. Pan, and D. P. Karl, "Energy Content of Eggs of the Brown Kiwi, *Apteryx australis*; an Extreme in Avian Evolution," *Comparative Biochemistry and Physiology Part A: Physiology* 60 (1978): 177–79.
5. L. W. Simmons, R. C. Firman, G. Rhodes, and M. Peters, "Human Sperm Competition: Testis Size, Sperm Production and Rates of Extrapair Copulations," *Animal Behaviour* 68 (2004): 297–302.
6. Another reason parental care is generally provided by females, rather than males, has to do with the certainty of parentage. For many animal species females retain the eggs inside their bodies until after they are fertilized. A female who cares for these eggs can be certain that they are hers, and not those of another female, so energy and time invested are well spent. Males have no such assurance, for precisely the same reason: if fertilization happens inside females then males run the risk that sperm from a rival male actually fertilizes the offspring. Expending resources for the care of unrelated offspring is not cost-effective. Consequently, because females have invested the most already, and because they generally have higher certainty of genetic parentage than their mates, selection favors the evolution of maternal care of offspring more often than it does paternal care.
7. Cockroaches actually show all sorts of interesting forms of parental care. A good review is provided by Christine Nalepa and William Bell, "Postovulation Parental Investment and Parental Care in Cockroaches," in *The Evolution of Social Behavior in Insects and Arachnids*, ed. Jae Choe and Bernard Crespi (Cambridge: Cambridge University Press, 1997), 26–51.
8. T. G. Benton, "Reproduction and Parental Care in the Scorpion, *Euscorpius flavicaudis*," *Behaviour* 117 (1991): 20–29.
9. G. Halffter and W. D. Edmonds, *The Nesting Behavior of Dung Beetles (Scarabaeinae): An Ecological and Evolutive Approach* (Mexico, D.F.: Instituto de Ecologia, 1982).
10. The concept described here is called the "operational sex ratio" (OSR). Whereas the sex

ratio is simply a head count of the number of males and females in a population (and, in all but a very few exceptional species, this ratio hovers near 1:1), the operational sex ratio accounts for the fact that not all individuals are actually available for reproducing at any point in time. It is defined as the ratio of reproductively available males to reproductively available females. The OSR can skew in the direction of females, as it does in jacanas, but it is typically skewed toward an excess of available males. The extent of skew is a good metric for the intensity of sexual selection likely to be acting in the population. The foundational paper describing this concept was written by my father, Stephen Emlen, and Lewis Oring, "Ecology, Sexual Selection, and the Evolution of Mating Systems," *Science* 197 (1977): 215–23. A more recent twist on these concepts is provided by H. Kokko and P. Monaghan, "Predicting the Direction of Sexual Selection," *Ecology Letters* 4 (2001): 159–65.

11. C. Darwin, *The Descent of Man and Selection in Relation to Sex* (London: John Murray, 1871).
12. S. T. Emlen and P. H. Wrege, "Size Dimorphism, Intrasexual Competition, and Sexual Selection in the Wattled Jacana (*Jacana jacana*), a Sex-Role-Reversed Shorebird in Panama," *Auk* 121 (2004): 391–403; Emlen and Wrege, "Division of Labor in Parental Care Behavior of a Sex-Role-Reversed Shorebird, the Wattled Jacana," *Animal Behaviour* 68 (2004): 847–55.
13. If you're wondering why it's the male who cares for the young in this species, that's a question my dad spent all those years answering, and a topic for a different book. But I can refer you to his paper: Emlen and Wrege, "Division of Labor in Parental Care Behavior," 847–55.
14. J. H. Poole, "Mate Guarding, Reproductive Success, and Female Choice in African Elephants," *Animal Behaviour* 37 (1989): 842–49.
15. Ibid.
16. J. A. Hollister-Smith, J. H. Poole, E. A. Archie, E. A. Vance, N. J. Georgiadis, C. J. Moss, and S. C. Alberts, "Age, Musth, and Paternity Success in Wild Male African Elephants, *Loxodonta Africana*," *Animal Behaviour* 74 (2006): 287–96.
17. H. F. Osborn, "The Ancestral Tree of the Proboscidea: Discovery, Evolution, Migration, and Extinction over a 50,000,000 Year Period," *Proceedings of the National Academy of Sciences* 21 (1935): 404–12; J. Shoshani and T. Pascal, eds., *The Proboscidea: Evolution and Paleoecology of Elephants and Their Relatives* (Oxford: Oxford University Press, 1993); W. J. Sanders, "Proboscidea," in *Paleontology and Geology of Laetoli: Human Evolution in Context*, vol. 2, *Fossil Hominins and the Associated Fauna*, ed. T. Harrison (New York: Springer, 2011).
18. F. Kottenkamp, *History of Chivalry and Ancient Armour* (London: Willis and Sotheran, 1857); G. Duby, *The Chivalrous Society*, trans. Cynthia Poston (Berkeley: University of California Press, 1977); R. L. O'Connell, *Of Arms and Men: A History of War, Weapons, and Aggression* (Oxford: Oxford University Press, 1989); J. France, *Western Warfare in the Age of the Crusades, 1000–1300* (Ithaca, NY: Cornell University Press, 1999).
19. Duby, *Chivalrous Society.*
20. Ibid.
21. Ibid.
22. Ibid.
23. Ibid.
24. Ibid.
25. Ibid.
26. Ibid.
27. Ibid.
28. Duby, *Chivalrous Society*; O'Connell, *Of Arms and Men.*
29. O'Connell, *Of Arms and Men.*
30. Duby, *Chivalrous Society*; O'Connell, *Of Arms and Men.*
31. Kottenkamp, *History of Chivalry and Ancient Armour*; O'Connell, *Of Arms and Men.*
32. Reproduction outside of marriage was also tied with rank and wealth. To an extraordinary degree, powerful lords sequestered young women inside the walls of their castles, plucking them from their homesteads to work as maids and attendants. There is abundant evidence

that these lords mated prolifically with harems of these women, often siring dozens of illegitimate children. In some cases, they prevented these women from marrying other men; in others, they sequestered them as virgins, and then married them off after they had had a child with the castle lord. For a comprehensive treatment of these issues, I recommend the writings of Laura Betzig, including "Medieval Monogamy," *Journal of Family History* 20 (1995): 181–216; and *Despotism and Differential Reproduction: A Darwinian View of History* (Hawthorne, NY: Aldine, 1986).

33. Michael Bell, Jeffrey Baumgartner, and Everett Olson, "Patterns of Temporal Change in Single Morphological Characters of a Miocene Stickleback Fish," *Paleobiology* 11 (1985): 258–71.
34. A fantastic illustration of this is the long-term study of bill evolution in Galapagos finches conducted by Peter and Rosemary Grant. For more than forty years they measured natural selection and evolution of beak shape in a population of seed-eating ground finches living on the tiny island of Daphne Major. Birds with thick beaks were better at cracking large, tough seeds, while those with slender beaks were faster at feeding on tiny seeds. The Grants showed that year-to-year fluctuations in rainfall led to dramatic shifts in the types and amounts of seeds available to the birds, and this resulted in selection favoring deep beaks in some years but thin beaks in others. Although natural selection was directional and strong for most years of this period, the pattern of selection oscillated, so that the net effect was stasis. Despite multiple bouts of rapid change, birds at the end of the sample period had roughly the same bill shapes as those at the beginning. The first thirty years of this study are described in Peter Grant and Rosemary Grant, "Unpredictable Evolution in a 30-Year Study of Darwin's Finches," *Science* 296 (2002): 707–11.
35. Charles Darwin first recognized this special property of sexual selection in *The Descent of Man and Selection in Relation to Sex* (London: John Murray, 1871), but the logic of escalation and unending change was best (and beautifully) articulated in two papers by Mary Jane West Eberhard, "Sexual Selection, Social Competition, and Evolution," *Proceedings of the American Philosophical Society* 123 (1979): 222–34; and "Sexual Selection, Social Competition, and Speciation," *Quarterly Review of Biology* 58 (1983): 155–83.
36. I should add the caveat "adult" animals, as in many species juvenile mortality is severe, and selection for traits that facilitate survival to the age of reproduction is just as strong, or even stronger, than selection for traits involved in reproduction. For an example of this, see the studies of natural and sexual selection in water striders by Daphne Fairbairn, such as R. F. Preziosi and D. J. Fairbairn, "Lifetime Selection on Adult Body Size and Components of Body Size in a Waterstrider: Opposing Selection and Maintenance of Sexual Size Dimorphism," *Evolution* 54 (2000): 558–66.

5. Economic Defensibility

1. Much of the work on túngara frogs has been conducted by Michael Ryan. See, for example, his book *The Túngara Frog: A Study in Sexual Selection and Communication* (Chicago, University of Chicago Press, 1992); or his paper "Female Mate Choice in a Neotropical Frog," *Science* 209 (1980): 523–25.
2. An absolutely gorgeous book on birds of paradise is Tim Layman and Edwin Scholes, *Birds of Paradise: Revealing the World's Most Extraordinary Birds* (Washington, DC: National Geographic, 2012). Early studies of female choice in these birds were conducted by Stephen Pruett-Jones and his students. See, for example, S. G. Pruett-Jones and M. A. Pruett-Jones, "Sexual Selection Through Female Choice in Lawes' Parotia, a Lek-Mating Bird of Paradise," *Evolution* 44 (1990): 486–501.
3. The concept of female choice traces to Darwin (*The Descent of Man and Selection in Relation to Sex*, London: John Murray, 1871), and this aspect of sexual selection has been the focus of literally thousands of empirical studies in all sorts of interesting species. A good place to start delving into this topic is the overview of sexual selection provided by Malte Andersson, *Sexual Selection* (Princeton, NJ: Princeton University Press, 1994).
4. My colleagues and I used information from the DNA sequences of approximately fifty spe-

cies of the dung beetle genus *Onthophagus* to arrange taxa into a nested series of groups based on their relatedness. The resulting tree, called a phylogeny, describes the history of these animals and can be used to trace the evolution of particular traits such as horns. This study showed that horns were gained and lost repeatedly in the history of these beetles. D. J. Emlen, J. Marangelo, B. Ball, and C. W. Cunningham, "Diversity in the Weapons of Sexual Selection: Horn Evolution in the Beetle Genus *Onthophagus* (Coleoptera: Scarabaeidae)," *Evolution* 59 (2005): 1060–84.

5. Ilkka Hanski and Yves Cambefort, eds., *Dung Beetle Ecology* (Princeton, NJ: Princeton University Press, 1991), provides a readable and comprehensive account of the geographical distribution of dung beetles, as well as their evolutionary history.
6. A great treatment of these topics is provided in chapters 8 and 11 of John Alcock, *Animal Behavior*, 8th ed. (Sunderland, MA: Sinauer Associates, 2005). A classic paper applying these concepts to sexual selection and the evolution of animal behavior is the paper by my father, Stephen Emlen, and Lewis Oring, "Ecology, Sexual Selection, and the Evolution of Mating Systems," *Science* 197 (1977): 215–23.
7. Jeanne and David Zeh describe their adventures in Panamanian forests in search of harlequin beetles in "Tropical Liaisons on a Beetle's Back," *Natural History* (1994): 36–43. Their results are published in the article "Sexual Selection and Sexual Dimorphism in the Harlequin Beetle *Acrocinus longimanus*," *Biotropica* 24 (1992): 86–96.
8. The Zehs' work on pseudoscorpions is published in the papers "Dispersal-Generated Sexual Selection in a Beetle-Riding Pseudoscorpion," *Behavioral Ecology and Sociobiology* 30 (1992): 135–42, and "Sex Via the Substrate: Sexual Selection and Mating Systems in Pseudoscorpions," in *The Evolution of Mating Systems in Insects and Arachnids*, ed. J. C. Choe and B. J. Crespi (Cambridge: Cambridge University Press, 1997): 329–39. Much of their recent work focuses on these tiny arthropods. Now that they have developed methods for rearing them in captivity in their laboratory, the Zehs have looked extensively at the genetic benefits that female pseudoscorpions derive from mating with dominant males on the backs of beetles, and from mating with several different males as they ride on beetle after beetle. See, for example, their paper "Genetic Benefits Enhance the Reproductive Success of Polyandrous Females," *Proceedings of the National Academy of Sciences* 96 (1999): 10236–41.
9. Wonderful accounts of the behavior of dung beetles are provided in Gonzalo Halffter and Eric G. Matthews, *The Natural History of Dung Beetles of the Subfamily Scarabaeinae (Coleoptera, Scarabaeidae)* (Palermo, Italy: Medical Books di G. Cafaro, 1966); Gonzalo Halffter and William David Edmonds, *The Nesting Behavior of Dung Beetles (Scarabaeinae): An Ecological and Evolutive Approach* (Mexico, D.F.: Instituto de Ecologia, 1982); and Leigh W. Simmons and James T. Ridsdill-Smith, *Ecology and Evolution of Dung Beetles* (Oxford: Blackwell Publishing, 2011). I also recommend papers by Hiroaki Sato, such as H. Sato and M. Imamori, "Nesting Behaviour of a Subsocial African Ball-Roller *Kheper platynotus* (Coleoptera, Scarabaeidae)," *Ecological Entomology* 12 (1987): 415–25; and H. Sato, "Two Nesting Behaviours and Life History of a Subsocial African Dung Rolling Beetle, *Scarabaeus catenatus* (Coleoptera: Scarabaeidae)," *Journal of Natural History* 31 (1997): 457–69.
10. Keith Philips and I used a phylogenetic tree depicting the branching relationships among dung beetle species to test for an association between evolutionary gains or losses of male horns and the tunneling versus rolling behavior of each species. We found that tunneling behavior strongly predicted the evolution of horns, and when species switched from tunneling to ball-rolling behavior, they subsequently lost their male horns. D. J. Emlen and T. K. Philips, "Phylogenetic Evidence for an Association Between Tunneling Behavior and the Evolution of Horns in Dung Beetles (Coleoptera: Scarabaeidae: Scarabaeinae)," in *Coleopterists Society Monographs* 5 (2006): 47–56.
11. D. J. Emlen, "Alternative Reproductive Tactics and Male Dimorphism in the Horned Beetle *Onthophagus acuminatus*," *Behavioral Ecology and Sociobiology* 41 (1997): 335–41; A. P. Moczek and D. J. Emlen, "Male Horn Dimorphism in the Scarab Beetle *Onthophagus taurus*: Do Alternative Tactics Favor Alternative Phenotypes?" *Animal Behaviour* 59 (2000): 459–66.

6. Duels

1. For a biography of Lanchester see P. W. Kingsford, *F. W. Lanchester: A Life of an Engineer* (London: Edward Arnold, 1960).
2. Frederick W. Lanchester, *Aircraft in Warfare: The Dawn of the Fourth Arm* (London: Constable, 1916).
3. Phillip M. Morse and George E. Kimball, *Methods of Operations Research* (New York: John Wiley and Sons, 1951); and James G. Taylor, *Lanchester Models of Warfare* (Arlington, VA: Operations Research Society of America, 1983).
4. For example, see P. R. Wallis, "Recent Developments in Lanchester Theory," *Operations Research* 19 (1968): 191–95, which reports on the Operational Research Conference on Recent Developments in Lanchester Theory, sponsored by the NATO Science Committee and held in Munich in July 1967.
5. For examples, see P. Morse and G. Kimball, *Methods of Operations Research* (Cambridge, MA: Technology Press of MIT, 1951), or Frederick S. Hillier and Gerald J. Lieberman, *Introduction to Operations Research*, 9th ed. (Boston: McGraw Hill, 2009). In 1956, the journal *Operations Research* dedicated an issue to the memory of Frederick Lanchester: Joseph McCloskey, "Of Horseless Carriages, Flying Machines, and Operations Research: A Tribute to Frederick Lanchester," 4 (1956): 141–47. To this day, the Institute for Operations Research and Management Sciences (INFORMS) names its highest prize after Lanchester.
6. For a superb explanation of the logic of Lanchester's models see John W. R. Lepingwell, "The Laws of Combat? Lanchester Re-examined," *International Security* 12 (1987): 89–134.
7. John Keegan, *The Face of Battle: A Study of Agincourt, Waterloo, and the Somme* (London: Penguin Books, 1983).
8. Lanchester, *Aircraft in Warfare*; Lepingwell, "Laws of Combat?": 89–134.
9. In fact, at Agincourt, the advantage worked the other way. The French marched into battle using traditional close-range tactics of knights in armor, but the English combined the strength of their knights with a new type of weapon, the longbow, wielded by thousands of archers. The English were able to concentrate arrow fire in a way that the French could not, and, despite being drastically outnumbered at the outset of the battle, the English won the day. For reasons we come back to in later chapters, this battle and others like it (for example, the Battle of Crécy) marked significant turning points in the nature of battle, spelling the beginning of the end for knights resplendent in suits of expensive armor.
10. For examples, see J. H. Engel, "A Verification of Lanchester's Law," *Operations Research* 2 (1954): 163–71; Thomas W. Lucas and Turker Turkes, "Fitting Lanchester's Equations to the Battles of Kursk and Ardennes," *Naval Research Logistics* 54 (2003): 95–116; and Taylor, *Lanchester Models of Warfare*.
11. Lepingwell, "Laws of Combat?," 89–134.
12. For explanations of this logic and application to military history, see R. L. O'Connell, *Of Arms and Men: A History of War, Weapons, and Aggression* (Oxford: Oxford University Press, 1990).
13. Many social insects do fight battles as armies, with swarms from one colony engaging swarms from another. Several authors have applied Lanchester's linear and square laws to these battles. For example, see N. R. Franks and L. W. Partridge, "Lanchester Battles and the Evolution of Combat in Ants," *Animal Behaviour* 45 (1993): 197–99; T. P. McGlynn, "Do Lanchester's Laws of Combat Describe Competition in Ants?" *Behavioral Ecology* 11 (2000): 686–90; Martin Pfeiffer and Karl E. Linsenmair, "Territoriality in the Malaysian Giant Ant *Camponotus gigas* (Hymenoptera/Formicidae)," *Journal of Ethology* 19 (2001): 75–85; and Nicola J. R. Plowes and Eldridge S. Adams, "An Empirical Test of Lanchester's Square Law: Mortality During Battles of the Fire Ant *Solenopsis invicta*," *Behavioral Ecology* 272 (2005): 1809–14.
14. Jon M. Hastings, "The Influence of Size, Age, and Residency Status on Territory Defense in Male Western Cicada Killer Wasps (*Sphecius grandis*, Hymenoptera: Sphecidae)," *Journal of the Kansas Entomological Society* 62 (1989): 363–73.
15. Cicada-killer wasps may lack big weapons for a second reason, too. For many insects that fight in the air, agility and maneuverability matter even more than strength or size. Big weapons

may hinder mobility in these fights in much the same way that weapons impede speed in predators. Many wasps, dragonflies, damselflies, and butterflies fight vicious, acrobatic airborne battles, and almost all of these species lack elaborate weapons. Part of this, no doubt, is due to unpredictability arising from scrambles; the rest is likely balancing selection arising from the need for agility. For example papers of fights in these types of insects, see Greg F. Grether, "Intrasexual Competition Alone Favors a Sexually Selected Dimorphic Ornament in the Rubyspot Damselfly *Hetaerina americana*," *Evolution* 50 (1996): 1949–57; D. J. Kemp and C. Wiklund, "Fighting Without Weaponry: A Review of Male-Male Contest Competition in Butterflies," *Behavioral Ecology and Sociobiology* 49 (2001): 429–42; J. Contreras-Garduño, J. Canales-Lazcana, and A. Córdoba-Aguilar, "Wing Pigmentation, Immune Ability, Fat Reserves, and Territorial Status in Males of the Rubyspot Damselfly, *Hetaerina americana*," *Journal of Ethology* 24 (2006): 165–73; M. A. Serrano-Meneses, A. Córdoba-Aguilar, V. Méndez, S. J. Layen, and T. Székely, "Sexual Size Dimorphism in the American Rubyspot: Male Body Size Predicts Male Competition and Mating Success," *Animal Behaviour* 73 (2007): 987–97.

16. The most comprehensive work on mating behavior and sexual selection in horseshoe crabs has been conducted by Jane Brockmann and her students at the University of Florida. For example, see H. Jane Brockmann and Dustin Penn, "Male Mating Tactics in the Horseshoe Crab, *Limulus polyphemus*," *Animal Behaviour* 44 (1992): 653–65.
17. O'Connell, *Of Arms and Men*.
18. J. H. Christy and M. Salmon, "Ecology and Evolution of Mating Systems of Fiddler Crabs (Genus *Uca*)," *Biological Reviews* 59 (1984): 483–509; N. Knowlton and B. D. Keller, "Symmetric Fights as a Measure of Escalation Potential in a Symbiotic, Territorial Snapping Shrimp," *Behavioral Ecology and Sociobiology* 10 (1982): 289–92; M. D. Jennions and P. R. Y. Backwell, "Residency and Size Affect Fight Duration and Outcome in the Fiddler Crab *Uca annulipes*," *Biological Journal of the Linnean Society* 57 (1996): 293–306.
19. The most comprehensive work to date on these bizarre wasps has been done by Robert Longair of the University of Calgary. His field studies in the Ivory Coast showed that males used their long tusks in fights with rival males over mud burrow–like nests on the undersides of leaves containing newly emerging females. See, for example, Robert W. Longair, "Tusked Males, Male Dimorphism, and Nesting Behavior in a Subsocial Afrotropical Wasp, *Synagris cornuta*, and Weapons and Dimorphism in the Genus (Hymenoptera: Vespidae: Eumeninae)," *Journal of the Kansas Entomological Society* 77 (2004): 528–57.
20. An early study from 1931 showed that the rhinoceros beetle *Diloboderus* fights over burrows in the soil: J. B. Daguerre, "Costumbres Nupciales del *Diloboderus abderus* Sturm," *Rev. Soc. Entomologia Argentina* 3 (1931): 253–56. Several studies by William Eberhard have looked at fighting behavior of rhinoceros beetles inside hollowed-out plant stems. See his book chapter "The Function of Horns in *Podischnus agenor* (Dynastinae) and Other Beetles" in *Sexual Selection and Reproductive Competition in Insects*, ed. M. S. Blum and N. A. Blum (New York: Academic Press, 1979), 231–59, and his paper "Use of Horns in Fights by the Dimorphic Males of *Ageopsis nigricollis* Coleoptera Scarabeidae, Dynastinae," *Journal of the Kansas Entomological Society* 60 (1987): 504–9.
21. Tusked frogs are a truly bizarre group of amphibian species. For more on their morphology and behavior I particularly recommend Sharon Emerson, "Courtship and Nest-Building Behavior of a Bornean Frog, *Rana blythi*," *Copeia* 1992 (1992): 1123–27; Kaliope Katsikaros and Richard Shine, "Sexual Dimorphism in the Tusked Frog, *Adelotus brevis* (Anura: Myobatrachidae): the Roles of Natural and Sexual Selection," *Biological Journal of the Linnean Society* 60 (1997): 39–51; and Hiroshi Tsuji and Masafumi Matsui, "Male-Male Combat and Head Morphology in a Fanged Frog (*Rana kuhlii*) from Taiwan," *Journal of Herpetology* 36 (2002): 520–26.
22. S. S. B. Hopkins, "The Evolution of Fossoriality and the Adaptive Role of Horns in the Mylagaulidae (Mammalia: Rodentia)," *Proceedings of the Royal Society of London Series B, Biological Sciences* 272 (2005): 1705–13.
23. My first attempt at dissertation research (before I ended up in Panama) was a study of horns

in the giant rhinoceros beetle *Golofa porteri* in Ecuador and southern Colombia, where these beetles fight to defend new plant shoots of a bamboo-like plant high up on cloud forest ridges. My attempts failed miserably, as I was unable to locate large populations to study. But that project was inspired by the beautiful paper by William Eberhard entitled "Fighting Behavior of Male *Golofa porteri* Beetles (Scarabaeidae: Dynastinae)," *Psyche* 83 (1978): 292–98.

24. Most work on the evolution of enlarged male hind legs in leaf-footed bugs has been done by Takahisa Miyatake of the University of Ryukyus in Japan. For example, see his paper "Territorial Mating Aggregation in the Bamboo Bug, *Notobitus meleagris*, Fabricius (Heteroptera: Coreidae)," *Journal of Ethology*, 13 (1995): 185–89, or his paper "Functional Morphology of the Hind Legs as Weapons for Male Contests in *Leptoglossus australis* (Heteroptera: Coreidae)," *Journal of Insect Behavior* 10 (1997): 727–35. Also see the paper by William Eberhard, "Sexual Behavior of *Acanthocephala declivis guatemalana* (Hemiptera: Coreidae) and the Allometric Scaling of their Modified Hind Legs," *Annals of the Entomological Society of America* 91 (1998): 863–71.

25. Horned chameleons, despite their notoriety, remain almost completely unstudied in the wild. One early study was conducted by Stanley Rand, "A Suggested Function of the Ornamentation of East African Forest Chameleons," *Copeia* 1961 (1961): 411–14. Another was conducted by Stephen Parcher, "Observations on the Natural Histories of Six Malagasy Chamaeleontidae," *Zeitschrift für Tierpsychologie* 34 (1974): 500–23.

26. Tadatsugu Hosoya and Kunio Araya have a beautiful paper entitled "Phylogeny of Japanese Stag Beetles (Coleoptera: Lucanidae) Inferred from 16S mtrRNA Gene Sequences, with Reference to the Evolution of Sexual Dimorphism of Mandibles," *Zoological Science* 22 (2005): 1305–18. In this paper the authors trace the evolutionary history of enlarged male mandibles in stag beetles. Their data suggest that huge weapons arose independently in at least two separate lineages of these beetles and that, once there, enlarged male mandibles were lost several different times. They are able to interpret these evolutionary losses of male weapons in the context of the natural history and behavior of the beetles.

27. These little flies have proven to be very difficult to study. They have yet to be successfully reared in the laboratory, and most species live in remote parts of New Guinea and surrounding islands. One species makes it into tropical northern Australia. Systematic study and taxonomy of these amazing flies has been conducted largely by David McAlpine, including "A Systematic Study of *Phytalmia* (Diptera, Tephritidae) with Description of a New Genus," *Systematic Entomology* 3 (1978): 159–75. The first field study of these flies that I am aware of is the paper by M. S. Moulds, "Field Observations on the Behavior of a North Queensland Species of *Phytalmia* (Diptera: Tephritidae)," *Journal of the Australian Entomological Society* 16 (1978): 347–52. More recently, the behavior of these flies has been studied by Gary Dodson, "Resource Defense Mating System in Antlered Flies, *Phytalmia spp.* (Diptera: Tephritidae)," *Annals of the Entomological Society of America* 90 (1997): 496–504.

28. The classic papers on stalk-eyed fly behavior are by Dietrich Burkhardt and Ingrid de la Motte, including their papers "Big 'Antlers' are Favoured: Female Choice in Stalk-Eyed Flies (Diptera, Insecta), Field Collected Harems and Laboratory Experiments," *Journal of Comparative Physiology A* 162 (1988): 649–52; and "Signalling Fitness: Larger Males Sire More Offspring: Studies of the Stalk-Eyed Fly *Cyrtodiopsis whitei* (Diopsidae, Diptera)," *Journal of Comparative Physiology A* 174 (1994): 61–4.

29. Gerald Wilkinson at the University of Maryland has studied the behavior and genetics of stalk-eyed flies for almost twenty years. He and many of his doctoral students and postdoctoral research fellows have constructed phylogenies for this family of flies, conducted multigeneration experiments with flies in the laboratory, and studied several species in the field. Some of my favorites of their papers are Patrick Lorch, Gerald Wilkinson, and Paul Reillo, "Copulation Duration and Sperm Precedence in the Stalk-Eyed Fly, *Cyrtodiopsis whitei* (Diptera: Diopsidae)," *Behavioral Ecology and Sociobiology* 32 (1993): 303–11; Gerald Wilkinson and Gary Dodson, "Function and Evolution of Antlers and Eye Stalks in Flies," in *The Evolu-*

tion of Mating Systems in Insects and Arachnids, ed. J. Choe and B. Crespi (Cambridge: Cambridge University Press, 1997), 310–28; and Tami Panhuis and Gerald Wilkinson, "Exaggerated Male Eye Span Influences Contest Outcome in Stalk-Eyed Flies," *Behavioral Ecology and Sociobiology* 46 (1999): 221–27. I also recommend Rick Baker and Gerald Wilkinson, "Phylogenetic Analysis of Eye Stalk Allometry and Sexual Dimorphism in Stalk-Eyed Flies (Diopsidae)," *Evolution* 55 (2001): 1373–85.

30. Kevin Fowler and Andrew Pomiankowski head a stalk-eyed fly research group at the University College London. Their group has combined field studies of sexual selection in these flies with laboratory studies of eyestalk development. I recommend Patrice David, Andrew Hingle, D. Greig, A. Rutherford, Andrew Pomiankowski, and Kevin Fowler, "Male Sexual Ornament Size but not Asymmetry Reflects Condition in Stalk-Eyed Flies," *Proceedings of the Royal Society B: Biological Sciences* 265 (1998): 2211–16; Andrew Hingle, Kevin Fowler, and Andrew Pomiankowski, "Size-Dependent Mate Preference in the Stalk-Eyed Fly *Cyrtodiopsis dalmanni*," *Animal Behaviour* 61 (2001): 589–95; and Jen Small, Sam Cotton, Kevin Fowler, and Andrew Pomiankowski, "Male Eyespan and Resource Ownership Affect Contest Outcome in the Stalk-Eyed Fly, *Teleopsis dalmanni*," *Animal Behaviour* 78 (2009): 1213–20.
31. For eloquent treatments of this spectacular period of naval warfare, see W. Murray, *The Age of the Titans: The Rise and Fall of the Great Hellenistic Navies* (Oxford: Oxford University Press, 2012); and John D. Grainger, *Hellenistic and Roman Naval Wars 336BC–31BC* (South Yorkshire, UK: Pen and Sword Books, 2011). The classic book on this period is Lionel Casson's *Ships and Seamanship in the Ancient World* (Baltimore: Johns Hopkins University Press, 1995), and I also recommend his book *The Ancient Mariners*, 2nd ed. (Princeton, NJ: Princeton University Press, 1991). For a particularly well-illustrated account of the ships and this period see Robert Gardiner, ed., *The Age of the Galley: Mediterranean Oared Vessels Since Pre-Classical Times* (London: Book Sales Publishing, 2000).
32. John Morrison and John Coates, *Greek and Roman Oared Warships 399–30BC* (Oxford: Oxbow Books, 1997).
33. My favorite account of this arms race, including especially the idea that the battering ram changed the nature of combat by causing ships to interact like individuals, thus fulfilling Lanchester's linear law, is provided by Robert L. O'Connell in his book *Of Arms and Men*, and in his book *Soul of the Sword: An Illustrated History of Weaponry and Warfare from Prehistory to the Present* (New York: Free Press, 2002).
34. John Morrison and John Coates, *Greek and Roman Oared Warships.*
35. John Morrison and Roderick Williams, *Greek Oared Ships* (Cambridge: Cambridge University Press, 1968).
36. Gardiner, *Age of the Galley.*
37. Casson, *Ships and Seamanship in the Ancient World*; Gardiner, *Age of the Galley*; and O'Connell, *Soul of the Sword.*
38. For information about the biology of mammoths I recommend Adrian Lister and Paul Bahn, *Mammoths: Giants of the Ice Age* (Berkeley: University of California Press, 2009). For a broader treatment of the extraordinary beasts of this age, I recommend Ian Lange, *Ice Age Mammals of North America : A Guide to the Big, the Hairy, and the Bizarre* (Missoula, MT: Mountain Press, 2002). And for an exciting narrative account of adventure in search of frozen mammoths, read Richard Stone, *Mammoth: The Resurrection of An Ice Age Giant* (Cambridge, MA: Perseus, 2002). For a comprehensive and technical treatment of fossil elephants and their relatives, I recommend J. Shoshani and P. Tassy, eds., *The Proboscidea: Evolution and Paleoecology of Elephants and Their Relatives* (Oxford: Oxford University Press, 1996).
39. I recommend the beautiful treatise on stag beetles by T. Mizunuma and S. Nagai, *The Lucanid Beetles of the World*, part of Mushi Sha's Iconographic Series of Insects, 1st ed., ed. H. Fijita (Tokyo: Mushi-Sha publishers, 1994). This book is written in Japanese, but has an English summary, and the illustrations speak for themselves!

40. Tadatsugu Hosoya and Kunio Araya, "Phylogeny of Japanese Stag Beetles (Coleoptera: Lucanidae) Inferred from 16s mtrRNA Gene Sequences, with References to the Evolution of Sexual Dimorphism of Mandibles," *Zoological Science* 22 (2005): 1305–18.
41. David Grimaldi and Gene Fenster, "Evolution of Extreme Sexual Dimorphisms: Structural and Behavioral Convergence Among Broad-Headed Male Drosophilidae (Diptera)," *American Museum Novitates* 2939 (1989): 1–25.
42. A comprehensive and well-written treatise on fossil ungulates and ungulate evolution is Donald Prothero and Robert Schoch, *Horns, Tusks, and Flippers: The Evolution of Hoofed Mammals* (Baltimore: Johns Hopkins University Press, 2003). I also recommend the more recent volume edited by Donald Prothero and Scott Foss, *The Evolution of Artiodactyls* (Baltimore: Johns Hopkins University Press, 2007); and the book by Elizabeth Vrba and George Schaller, eds., *Antelopes, Deer, and Relatives: Fossil Record, Behavioral Ecology, Systematics, and Conservation* (New Haven, CT: Yale University Press, 2000). All of these cover the major patterns of evolution (and weapon radiation) in ungulates. Several authors have also used diversity in behavior and morphology of living ungulates to suggest how and why weapons have diversified. For example, see the book by Valerius Geist, *Deer of the World: Their Evolution, Behaviour, and Ecology* (Mechanicsburg, PA: Stackpole Books, 1998); and these papers: T. M. Caro, C. M. Graham, C. J. Stoner, and M. M. Flores, "Correlates of Horn and Antler Shape in Bovids and Cervids," *Behavioral Ecology and Sociobiology* 55 (2003): 32–41; G. A. Lincoln, "Teeth, Horns and Antlers: The Weapons of Sex," in *The Differences Between the Sexes*, ed. R. V. Short and E. Balaban (Cambridge: Cambridge University Press, 1994): 131–58; J. Bro-Jørgensen, "The Intensity of Sexual Selection Predicts Weapon Size in Male Bovids," *Evolution* 61 (2007): 1316–26; and B. Lundrigan, "Morphology of Horns and Fighting Behavior in the Family Bovidae," *Journal of Mammology* 77 (1996): 462–75.

7. Costs

1. The beetle species I studied is a tiny brown dung beetle called *Onthophagus acuminatus*. They are active primarily in the early morning, but also throughout the daytime, and they specialize on dung from howler monkeys. They fly only inches above the leaf litter of the forest floor, hovering in gentle back-and-forth undulations as they clumsily make their way toward fresh dung. Once they find the dung, they burrow inside it and into the soil below (one reason they are almost never noticed).
2. All those forays into the forest were not without adventure. I remember one morning with frightful clarity. It had been one of the dreaded days when the monkeys were not at all close to my room. I had not heard the dawn chorus and, as a result, I'd had to hike three miles into the forest in search of dung. My first finds had all been tiny—not nearly enough to feed the hundreds of beetles waiting in their tubes in the lab. The problem with "false starts" such as these is that at each collection I use one of the pairs of surgical gloves I kept stashed in my pocket. After several insufficient stops I begin to run out of gloves. Mucking with monkey dung is unpleasant enough as it is; having disposable gloves gives me at least a modicum of cleanliness. This way, when I'm done, I can strip my gloves, turning them inside out and trapping the filth inside them as I go. This leaves my hands relatively clean and keeps dung out of my lunch, water bottle, and binoculars.

   Three miles from the lab and well into my last pair of gloves, I finally found the stash of dung that I needed. However, as I was filling my bags, I brushed against a palm frond that was covered with ticks. Any tropical biologist can tell you about "seed ticks," the bundles of babies (technically nymphs) that clump together on the tips of big leaves. Hundreds can pile together into a ball the size of a marble. Bump them and they hurl themselves onto your body and begin to disperse. For this reason, standard forest garb is long pants tucked into socks, and a loose-fitting long-sleeved cotton shirt tucked into the pants. This works pretty well to keep stinging ants or tiny ticks from getting directly onto your skin. The other trick is to carry tape. Masking tape is best, and for easy access most of us would stick several strips along our pants legs over our thighs. When we happened upon a ball of ticks we could grab the tape and use it to blot the babies off of us in seconds, before they got past our clothing barrier.

My dilemma that morning was that I was on my last pair of gloves, and they had already started to tear. If I peeled them off I would never be able to get them back on again. I couldn't manage the tape with the gloves on, and I didn't want to harvest all the dung bare-handed, so I ignored the ticks for the moment and finished filling my bags. It took only another five minutes or so before I was done but, to my horror, that was too long. I ripped the gloves off, grabbed the tape, and looked for the ticks. They were gone. In those few short minutes they had dispersed across my body, and I couldn't find them.

The three-mile run back to the lab was agony. I could feel the ticks crawling inside my shirt, down my legs and crotch, in my hair, behind my ears, even in my nose and around my eyes as I ran. I flung the dung bags into the lab and dashed to my room, jumping out of my clothes into a scalding and soapy shower, to no avail. They would not wash off. It took more than an hour with tweezers under an intense bright light to peel the ticks from my skin, plucking them from the places they had lodged themselves and sticking them onto strips of tape like so many flecks of pepper. In the end I counted almost eight hundred ticks on those clogged strips of tape. Though it had only been an hour since I bumped that unfortunate frond, most of the ticks had already injected their anticoagulant salivary juices, and the itching that commenced was unreal. I chewed antihistamines for a week. Convinced that she would think I was exaggerating, I promptly sent the tick-laden strips of tape to my girlfriend (now wife) back at Duke, where she was at that time a doctoral student. You can imagine how well that went down. Apparently, field-biologist bravado is not the same thing as romance.

3. D. J. Emlen, "Artificial Selection on Horn Length-Body Size Allometry in the Horned Beetle *Onthophagus acuminatus*," *Evolution* 50 (1996): 1219–30.

4. Trade-offs among developing structures are widespread in insects, and they constitute a dramatic cost of growing elaborate weapons. But this particular cost—stunted growth of other structures—applies primarily to insects such as beetles, flies, ants, and bees. It does not apply to any animals that I am aware of outside of insects. The reason almost certainly has to do with the way these particular insects develop. Specifically, it has to do with when, during development, all of the various adult structures are formed.

Exaggerated weapons of sexual selection always grow at the end of development, around the time when males reach sexual maturity. Deer, elk, and moose all begin antler growth only after males are young adults. Elephants and boars grow their tusks after they are adults. Even shrimp and crabs begin the process of enlargement of their fighting claws around the time that their gonads mature. In all of these animals, bodies have already grown and organs, tissues, and appendages are already at or near their adult proportions, long before the weapons begin to grow.

Weapons cannot stunt the growth of other traits in these animals because all of the other structures are produced first. In contrast, beetles, bees, flies, and ants all undergo metamorphosis as they develop, and this means that they grow their adult body parts at the same time as their weapons. Simultaneous growth exposes these structures to the insidious effects of resource limitation and allocation trade-offs.

5. The most convincing ways to show this all involve some type of experimental perturbation to growing animals. When I selected for longer horns in dung beetles, increased horn growth led to reduced eye size. Using a hot needle to kill the cells of the developing horn in another dung beetle species produced adults with unusually large testes, implicating a resource allocation trade-off between weapons and testes. L. W. Simmons and D. J. Emlen, "Evolutionary Trade-Off Between Weapons and Testes," *Proceedings of the National Academy of Sciences* 103 (2006): 16346–51. A reverse experiment, ablating cells of the developing

genitalia, increased horn length in still another species. A. P. Moczek and H. F. Nijhout, "Trade-Offs During the Development of Primary and Secondary Sexual Traits in a Horned Beetle," *American Naturalist* 163 (2004): 184–91. In species after species, when males produce disproportionately large weapons other structures suffer, including traits critical for reproduction such as testes.

6. K. Kawano, "Horn and Wing Allometry and Male Dimorphism in Giant Rhinoceros Beetles (Coleoptera: Scarabaeidae) of Tropical Asia and America," *Annals of the Entomological Society of America* 88 (1995): 92–99.
7. K. Kawano, "Cost of Evolving Exaggerated Mandibles in Stag Beetles (Coleoptera: Lucanidae)," *Annals of the Entomological Society of America* 90 (1997): 453–61.
8. Catherine Fry, then a doctoral student at the University of Maryland, used topical applications of a developmental hormone called "juvenile hormone" to perturb the growth of male eyestalks. This hormone is known to regulate growth of a number of insect structures, including the distorted heads and weapons of soldier castes in termites and ants. When she painted synthetic juvenile hormone onto developing male larvae, they emerged as adults with disproportionately long eyestalks—she perturbed the relative size of the male weapon. But these males also had drastically reduced testes and, when these males later mated, they were able to transfer only two-thirds as many sperm as males not exposed to the hormone. C. Fry, "Juvenile Hormone Mediates a Trade-Off Between Primary and Secondary Sexual Traits in Stalk-Eyed Flies," *Evolution and Development* 8 (2006): 191–201. For her doctoral dissertation, Catherine Fry used two related species of stalk-eyed flies that differ in the relative sizes of their eyestalks to investigate whether eye-span exaggeration results in trade-offs with other body parts. In one species, *Cyrtodiopsis dalmanni*, males have exaggerated eye spans that are much longer than those of females. In the other, *C. quinqueguttata*, both sexes have approximately equal eyestalks, which are relatively unexaggerated in length. She used a variety of experimental approaches (including artificial selection, application of exogenous juvenile hormone, and diet manipulation) to perturb the relative length of the eye stalks. She showed that exaggerated eye span in male *C. dalmanni* is accompanied by a decrease in two other features of head morphology, eye bulb size and eye stalk width, as well as compromised testis growth and sperm production.
9. The Australian sweat bee *Lasioglossum hemichalceum* produces males with robust, expanded heads and jaws that dwarf the remainders of their tiny bodies (think lentils glued to the tips of rice grains, and you get the general idea). These top-heavy fighters have puny little wings and almost no wing muscles, and they cannot fly. Instead, they remain in their home nest after they emerge, fighting to the death with their brothers for the opportunity to mate with their newly emerging sisters. This same bee species also has smaller, more "typical" males that lack weapons, and these unarmed males can fly. They disperse from their burrow to seek out neighboring nests and females. After the big-headed males have mated with all of their sisters, they, too, venture out from their nest in search of females. But, unlike their flying counterparts, these males must crawl. Crawling the tens of meters between nests is an exceedingly laborious process for a rice-grain-sized bee with a bloated head, and it is almost always fatal because they're exposed to predators all along the way. The price of winning fights for the chance to mate with your sisters is an utter inability to make it to neighboring nests. P. F. Kukuk and M. Schwarz, "Macrocephalic Male Bees as Functional Reproductives and Probable Guards," *Pan-Pacific Entomologist* 64 (1988): 131–37.
10. *Cardiocondyla* ants produce two types of male, one type that fights and another that does not. Here, too, males develop either with formidable weapons (enlarged heads and jaws), or with wings and wing musculature, but never both. J. Heinze, B. Hölldobler, and K. Yamauchi, "Male Competition in *Cardiocondyla* Ants," *Behavioral Ecology and Sociobiology* 42 (1998): 239–46; S. Cremer and J. Heinze, "Adaptive Production of Fighter Males: Queens of the Ant *Cardiocondyla* Adjust the Sex Ratio Under Local Mate Competition," *Proceedings of the Royal Society of London, Series B* 269 (2002): 417–22.

11. J. Crane, *Fiddler Crabs of the World (Ocypodidae: Genus Uca)* (Princeton, NJ: Princeton University Press, 1975).
12. B. J. Allen and J. S. Levinton, "Costs of Bearing a Sexually Selected Ornamental Weapon in a Fiddler Crab," *Functional Ecology* 21 (2007): 154–61.
13. Masatoshi Matsumasa and Minoru Murai were able to measure the energetic costs of fiddler crab claws in action by tracking changes in blood glucose (a sugar used to power activity) and blood lactate (a chemical by-product of metabolism and an indication of energy burned) as animals performed various behaviors. By measuring baseline lactate levels in resting animals, and comparing this to the elevated levels they detected during fights, Matsumasa and Murai showed that the energetic cost of waving a claw was substantial. M. Matsumasa and M. Murai, "Changes in Blood Glucose and Lactate Levels of Male Fiddler Crabs: Effects of Aggression and Claw Waving," *Animal Behaviour* 69 (2005): 569–77.
14. Allen and Levinton, "Costs of Bearing a Sexually Selected Ornamental Weapon," 154–61.
15. I. Valiela, D. F. Babiec, W. Atherton, S. Seitzinger, and C. Krebs, "Some Consequences of Sexual Dimorphism: Feeding in Male and Female Fiddler Crabs, *Uca pugnax* (Smith)," *Biological Bulletin* 147 (1974): 652–60.
16. H. E. Caravello and G. N. Cameron, "The Effects of Sexual Selection on the Foraging Behaviour of the Gulf Coast Fiddler Crab, *Uca panacea*," *Animal Behaviour* 35 (1987): 1864–74.
17. T. Koga, P. R. Y. Backwell, J. H. Christy, M. Murai, and E. Kasuya, "Male-Biased Predation of a Fiddler Crab," *Animal Behaviour* 62 (2007): 201–7.
18. M. E. Cummings, J. M. Jordão, T. W. Cronin, and R. F. Oliveira, "Visual Ecology of the Fiddler Crab, *Uca tangeri*: Effects of Sex, Viewer and Background on Conspicuousness," *Animal Behaviour* 75 (2008): 175–88.
19. J. M. Jordão and R. F. Oliveira, "Sex Differences in Predator Evasion in the Fiddler Crab *Uca tangeri* (Decapoda: Ocypodidae)," *Journal of Crustacean Biology* 21 (2001): 948–53.
20. T. Koga et al., "Male-Biased Predation of a Fiddler Crab," 201–7. Predation isn't always male biased, however. For example, another study found that ibis preferred female fiddlers over males, perhaps because big claws made males harder to swallow. Keith Bildstein, Susan G. McDowell, and I. Lehr Brisbin, "Consequences of Sexual Dimorphism in Sand Fiddler Crabs, *Uca pugilator*: Differential Vulnerability to Avian Predation," *Animal Behaviour* 37 (1989): 133–39.
21. A. G. McElligott and T. J. Hayden, "Lifetime Mating Success, Sexual Selection and Life History of Fallow Bucks (*Dama dama*)," *Behavioral Ecology and Sociobiology* 48 (2000): 203–10.
22. R. Moen, J. Pastor, and Y. Cohen, "A Spatially Explicit Model of Moose Foraging and Energetics," *Ecology* 78 (1997): 505–21.
23. R. Moen and J. Pastor, "A Model to Predict Nutritional Requirements for Antler Growth in Moose," *Alces* 34 (1998): 59–74.
24. A. Bubenik, "Evolution, Taxonomy, and Morphophysiology," in *Ecology and Management of the North American Moose*, eds. A. W. Franzmann and C. C. Schwartz (University Press of Colorado, 2007): 77–123.
25. T. H. Clutton-Brock, "The Functions of Antlers," *Behaviour* 79 (1982): 108–24.
26. R. Moen, J. Pastor, and Y. Cohen, "Antler Growth and Extinction of Irish Elk," *Evolutionary Ecology Research* 1 (1999): 235–49.

8. Reliable Signals

1. All exaggerated "sexually selected" animal structures, including flashy ornaments of male displays and extreme male weapons, delay growth until relatively late in an animal's development, shooting to full size after most or all of the rest of the body is already in place. For example, rapid claw growth in crabs begins at the final, puberty molt. R. G. Hartnoll, "Variations in Growth Pattern Between Some Secondary Sexual Characters in Crabs (Decapoda, Brachyura)," *Crustaceana* 27 (1974): 131–36; Pitchaimuthu Mariappan, Chellam Balasundaram, and Barbara Schmitz, "Decapod Crustacean Chelipeds: An Over-

view," *Journal of Bioscience* 25 (2000): 301–13. Antler growth begins at puberty, as does narwhal tusk growth and walrus tusk growth. G. A. Lincoln, "Teeth, Horns and Antlers: The Weapons of Sex," in *Differences Between the Sexes,* eds. R. V. Short and E. Balaban (Cambridge: Cambridge University Press, 1994): 131–58; H. B. Silverman and M. J. Dunbar, "Aggressive Tusk Use by the Narwhal (*Monodon monoceros* L.)," *Nature* 284 (1980): 57–58; Edward Miller, "Walrus Ethology. I. The Social Role of Tusks and Applications of Multidimensional Scaling," *Canadian Journal of Zoology* 53 (1975): 590–613.

2. Although I don't discuss this further in this book, females in many beetles, some decapods, and many of the armed dinosaurs, fish, and ungulates, also produce weapons. In virtually every case these female weapons are similar to, but smaller than, the corresponding structures of males. In the few studied examples, females use horns in fights with conspecific females, generally over food resources or to protect young. See, for example, N. Knowlton and B. D. Keller, "Symmetric Fights as a Measure of Escalation Potential in a Symbiotic, Territorial Snapping Shrimp," *Behavioral Ecology and Sociobiology* 10 (1982): 289–92; J. Berger and C. Cunningham, "Phenotypic Alterations, Evolutionarily Significant Structures, and Rhino Conservation," *Conservation Biology* 8 (1994): 833–40; and V. O. Ezenwa and A. E. Jolles, "Horns Honestly Advertise Parasite Infection in Male and Female African Buffalo," *Animal Behaviour* 75 (2008): 2013–21. Two comparative studies explicitly tested for a role of sexual selection in the evolution of female weapons, and concluded that these structures most likely had been shaped by natural, rather than sexual selection. T. M. Caro, C. M. Graham, C. J. Stoner, and M. M. Flores, "Correlates of Horn and Antler Shape in Bovids and Cervids," *Behaviorial Ecology and Sociobiology* 55 (2003): 32–41; J. Bro-Jørgensen, "The Intensity of Sexual Selection Predicts Weapon Size in Male Bovids," *Evolution* 61 (2007) 1316–26. Several reviews have focused on this topic. See, for example, C. Packer, "Sexual Dimorphism: The Horns of African Antelopes," *Science* 221 (1983): 1191–93; R. A. Kiltie, "Evolution and Function of Horns and Horn-Like Organs in Female Ungulates," *Biological Journal of the Linnean Society* 24 (1985): 299–320; and S. C. Roberts, "The Evolution of Hornedness in Female Ruminants," *Behaviour* 133 (1996): 399–442. But a great many taxa with female weapons have yet to be studied, and basic questions still linger. For example, under what circumstances do female weapons evolve? Do weapons arise in both sexes due to natural selection (for example, as a defense against predators) and then subsequently become co-opted as signals in males? Or do these weapons arise initially in males, and only in specific circumstances become co-opted by females?

3. This experiment is described in Douglas J. Emlen, Ian A. Warren, Annika Johns, Ian Dworkin, and Laura Corley Lavine, "A Mechanism of Extreme Growth and Reliable Signaling in Sexually Selected Ornaments and Weapons," *Science* 337 (2012): 860–64.

4. J. L. Tomkins, "Environmental and Genetic Determinants of the Male Forceps Length Dimorphism in the European Earwig *Forficula auricularia* L.," *Behavioral Ecology and Sociobiology* 47 (1999): 1–8.

5. P. David, A. Hingle, D. Greig, A. Rutherford, A. Pomiankowski, and K. Fowler, "Male Sexual Ornament Size but Not Asymmetry Reflects Condition in Stalk-Eyed Flies," *Proceedings of the Royal Society of London, Series B* 265 (1998): 2211–16; R. J. Knell, A. Fruhauf, and K. A. Norris, "Conditional Expression of a Sexually Selected Trait in the Stalk-Eyed Fly *Diasemopsis aethiopica*," *Ecological Entomology* 24 (1999): 323–28.

6. F. E. French, L. C. McEwen, N. D. Magruder, R. H. Ingram, and R. W. Swift, "Nutrient Requirements for Growth and Antler Development in the White-Tailed Deer," *Journal of Wildlife Management* 20 (1956): 221–32; W. Leslie Robinette, C. Harold Baer, Richard E. Pillmore, and C. Edward Knittle, "Effects of Nutritional Change on Captive Mule Deer," *Journal of Wildlife Management* 37 (1974): 312–26.

7. Studies of antler growth in red deer (basically, European elk) have begun to elucidate the developmental mechanisms coupling antler growth with nutrition, and the mechanisms they find are thrillingly similar to what we observe in beetle horns. Cells in growing tips of antlers are sensitive to signaling through the insulin/insulin-like growth factor (IGF)

pathway, a physiological mechanism that dials cell proliferation up or down depending on the nutritional state of an animal. For relevant studies, see J. M. Suttie, I. D. Corson, P. D. Gluckman, and P. F. Fennessy, "Insulin-Like Growth Factor 1, Growth and Body Composition in Red Deer Stags," *Animal Production* 53 (1991): 237–42; J. L. Elliott, J. M. Oldham, G. W. Asher, P. C. Molan, and J. J. Bass, "Effect of Testosterone on Binding of Insulin-Like Growth Factor-I (IGF-I) and IGF-II in Growing Antlers of Fallow Deer (*Dama dama*)," *Growth Regulation* 6 (1996): 214; J. R. Webster, I. D. Corson, R. P. Littlejohn, S. K. Martin, and J. M. Suttie, "The Roles of Photoperiod and Nutrition in the Seasonal Increases in Growth and Insulin-Like Growth Factor-1 Secretion in Male Red Deer," *Animal Science* 73 (2001): 305–11.

8. P. Fandos, "Factors Affecting Horn Growth in Male Spanish Ibex (*Capra pyrenaica*)," *Mammalia* 59 (1995): 229–35; M. Giacometti, R. Willing, and C. Defila, "Ambient Temperature in Spring Affects Horn Growth in Male Alpine Ibexes," *Journal of Mammalogy* 83 (2002): 245–51.
9. M. Mulvey and J. M. Aho, "Parasitism and Mate Competition: Liver Flukes in White-Tailed Deer," *Oikos* 66 (1993): 187–92. Several studies have also shown that parasites cause antlers to be less symmetrical, rather than simply shorter. For example, see Ivar Folstad, Per Arneberg, and Andrew J. Karte, "Antlers and Parasites," *Oecologia* 105 (1996): 556–58; Eystein Markusson and Ivar Folstad, "Reindeer Antlers: Visual Indicators of Individual Quality?" *Oecologia* 110 (1997): 501–7.
10. Ezenwa and Jolles, "Horns Honestly Advertise Parasite Infection," 2013–21.
11. B. W. Tucker, "On the Effects of an Epicaridan Parasite, *Gyge branchialis*, on *Upogebia littoralis*," *Quarterly Journal of Microscope Science* 74 (1930): 1–118; R. G. Hartnoll, "*Entionella monensis* sp. nov., an Entoniscis Parasite of the Crab *Eurynome aspera* (Pennant)," *Journal of the Marine Biology Association of the United Kingdom* 39 (1960): 101–7; T. Yamaguchi and H. Aratake, "Morphological Modifications Caused by *Sacculina polygenea* in *Hemigrapsus sanguineus* (De Haan) (Brachyura: Grapsidae)," *Crustacean Research* 26 (1997): 125–145; Mariappan, Balasundaram, and Schmitz, "Decapod Crustacean Chelipeds," 301–13.
12. A number of sources speak to the incredible cost of arms and armor of medieval knights, including F. Kottenkamp, *History of Chivalry and Ancient Armour* (London: Willis and Sotheran Publishers, 1857); G. Duby, *The Chivalrous Society*, trans. Cynthia Poston (Berkeley: University of California Press, 1977); R. L. O'Connell, *Of Arms and Men: A History of War, Weapons, and Aggression* (Oxford: Oxford University Press, 1989); J. France, *Western Warfare in the Age of the Crusades; 1000–1300* (Ithaca, NY: Cornell University Press, 1999); Constance Brittain Bouchard, *Knights: In History and Legend* (Lane Cove, Australia: Global Book Publishing, 2009).
13. O'Connell, *Of Arms and Men*; Bouchard, *Knights*.
14. Bouchard, *Knights*.
15. Ibid.
16. Duby, *Chivalrous Society*; O'Connell, *Of Arms and Men*; France, *Western Warfare in the Age of the Crusades*; Bouchard, *Knights*.
17. Ibid.
18. Ibid.
19. Hypervariability is a signature characteristic of the showiest, most exaggerated ornaments and weapons of sexual selection. For papers discussing this phenomenon in ornaments, see R. V. Alatalo, J. Höglund, and A. Lundberg, "Patterns of Variation in Tail Ornament Size in Birds," *Biological Journal of the Linnean Society of London* 34 (1988): 363; S. Fitzpatrick, "Patterns of Morphometric Variation in Birds' Tails: Length, Shape and Variability," *Biological Journal of the Linnean Society of London* 62 (1997): 145; J. J. Cuervo and A. P. Møller, "The Allometric Pattern of Sexually Size Dimorphic Feather Ornaments and Factors Affecting Allometry," *Journal of Evolutionary Biology* 22 (2009): 1503. For papers discussing hypervariability in weapons, see H. Frederik Nijhout and Douglas J. Emlen, "The Development and Evolution of Exaggerated Morphologies in Insects," *Annual Review of Entomology*

45 (2000): 661–708; Astrid Kodric-Brown, Richard M. Sibly, and James H. Brown, "The Allometry of Ornaments and Weapons," *Proceedings of the National Academy of Sciences* 103 (2006): 8733–38; Douglas J. Emlen, "The Evolution of Animal Weapons," *Annual Review of Ecology, Evolution, and Systematics* 39 (2008): 387–413.

20. Lots of theoretical models delve into the characteristics of honest signals in animal communication, especially with reference to ornaments and weapons of sexual selection. Early models proposing that the unusual variability of these traits would amplify subtle differences among males are provided in Oren Hasson, "Sexual Displays as Amplifiers: Practical Examples with an Emphasis on Feather Decorations," *Behavioral Ecology* 2 (1991): 189–97. Excellent overviews of signaling theory, and of the ingredients of honest signals of male quality, are offered by John Maynard Smith and David Harper, *Animal Signals* (Oxford: Oxford University Press, 2003); William A. Searcy and Stephen Nowicki, *The Evolution of Animal Communication: Reliability and Deception in Signaling Systems* (Princeton, NJ: Princeton University Press, 2010); and Jack W. Bradbury and Sandra L. Vehrencamp, *Principles of Animal Communication*, 2nd ed. (Sunderland, MA: Sinauer Associates, 2011).
21. A number of theoretical models also conclude that small, poor-quality males pay a steeper price for big ornaments or weapons and that, as a result, it's not cost-effective for them to invest in full-sized structures. For example, see Astrid Kodric-Brown and Jim H. Brown, "Truth in Advertising: The Kinds of Traits Favored by Sexual Selection," *American Naturalist* 124 (1984): 309–23; Nadav Nur and Oren Hasson, "Phenotypic Plasticity and the Handicap Principle," *Journal of Theoretical Biology* 110 (1984): 275–98; David W. Zeh and Jeanne A. Zeh, "Condition-Dependent Sex Ornaments and Field Tests of Sexual-Selection Theory," *American Naturalist* 132 (1988): 454–59; Russell Bonduriansky and Troy Day, "The Evolution of Static Allometry in Sexually Selected Traits," *Evolution* 57 (2003): 2450–58; Kodric-Brown, Sibly, and Brown, "The Allometry of Ornaments and Weapons," 8733–38.

9. Deterrence

1. The beach we hiked into is called Playa Naranjo, in Santa Rosa National Park.
2. Two papers resulting from John Christy's dissertation work on Devilfish Key are J. H. Christy, "Adaptive Significance of Reproductive Cycles in the Fiddler Crab *Uca pugilator*: a Hypothesis," *Science* 199 (1978): 453–55; and J. H. Christy, "Female Choice in the Resource-Defense Mating System of the Sand Fiddler Crab, *Uca pugilator*," *Behavioral Ecology and Sociobiology* 12 (1983): 169–80.
3. The specific type of interaction described here is called "sequential assessment," and foundational papers using game theory to model how and when assessment evolves include J. Maynard Smith, "The Theory of Games and the Evolution of Animal Conflicts," *Journal of Theoretical Biology* 47 (1974): 209–21; G. A. Parker, "Assessment Strategy and the Evolution of Animal Conflicts," *Journal of Theoretical Biology* 47 (1974): 223–43; J. Maynard Smith and G. Parker, "The Logic of Asymmetric Contests," *Animal Behaviour* 24 (1976): 159–65; M. Enquist and O. Leimar, "Evolution of Fighting Behaviour: Decision Rules and Assessment of Relative Strength," *Journal of Theoretical Biology* 102 (1983): 387–410. More recent overviews of animal signaling, including types of assessment, are provided by J. Maynard Smith and D. Harper, *Animal Signals* (Oxford: Oxford University Press, 2003); W. A. Searcy and S. Nowicki, *The Evolution of Animal Communication: Reliability and Deception in Signaling Systems* (Princeton, NJ: Princeton University Press, 2010); J. W. Bradbury and S. L. Vehrencamp, *Principles of Animal Communication*, 2nd ed. (Sunderland, MA: Sinauer Associates, 2011).
4. Numerous studies demonstrate that males with bigger claws win fights over burrows, including J. Crane, "Combat, Display and Ritualization in Fiddler Crabs (Ocypodidae, genus *Uca*)," *Philosophical Transactions of the Royal Society of London, Series B* 251 (1966): 459–72; G. W. Hyatt and M. Salmon, "Combat in the Fiddler Crabs *Uca pugilator* and *U. pugnax*: A Quantitative Analysis," *Behaviour* 65 (1978): 182–211; M. D. Jennions and

P. R. Backwell, "Residency and Size Affect Fight Duration and Outcome in the Fiddler Crab *Uca annulipes*," Biological Journal of the Linnean Society 57 (1996): 293–306; A. E. Pratt, D. K. McLain, and G. R. Lathrop, "The Assessment Game in Sand Fiddler Crab Contests for Breeding Burrows," *Animal Behaviour* 65 (2003): 945–55. In a fun study using strain gauges, Jeff Levinton and Michael Judge showed that males with bigger claws exert more powerful closing forces. J. S. Levinton and M. L. Judge, "The Relationship of Closing Force to Body Size for the Major Claw of *Uca pugnax* (Decapoda: Ocypodidae)," *Functional Ecology* 7 (1993): 339–45.

5. Descriptions of the behavior of sand fiddler crabs, including contests over burrows, are provided in J. H. Christy, "Burrow Structure and Use in the Sand Fiddler Crab, *Uca pugilator*," *Animal Behaviour* 30 (1982): 687–94; Christy, "Female Choice in the Resource-Defense Mating System"; M. Salmon and G. W. Hyatt, "Spatial and Temporal Aspects of Reproduction in North Carolina Fiddler Crabs (*Uca pugilator*)," *Journal of Experimental Marine Biology and Ecology* 70 (1983): 21–43; J. Christy and M. Salmon, "Ecology and Evolution of Mating Systems of Fiddler Crabs (genus *Uca*)," *Biological Reviews* (1984): 483–509.
6. Stages of fiddler crab fights are described in Crane, "Combat, Display and Ritualization in Fiddler Crabs," 459–72; Hyatt and Salmon, "Combat in the Fiddler Crabs," 182–211; Jennions and Backwell, "Residency and Size Affect Fight Duration and Outcome in the Fiddler Crab *Uca annulipes*," 293–306.
7. Hyatt and Salmon, "Combat in the Fiddler Crabs," 182–211.
8. Ibid.
9. Ibid.
10. Maynard Smith, "Theory of Games and the Evolution of Animal Conflicts," 209–21; Parker, "Assessment Strategy and the Evolution of Animal Conflicts," 223–43; Smith and Parker, "Logic of Asymmetric Contests," 159–65; Enquist and Leimar, "Evolution of Fighting Behaviour," 387–410.
11. Takahisa Miyatake, "Territorial Mating Aggregation in the Bamboo Bug, *Notobitus meleagris*, Fabricius (Heteroptera: Coreidae)," *Journal of Ethology* 13 (1995): 185–89; Miyatake, "Multi-Male Mating Aggregation in *Notobitus meleagris* (Hemiptera: Coreidae)," *Annals of the Entomological Society of America* 95 (2002): 340–44. For descriptions of similar behavior in additional species of Coreid bug, see Miyatake, "Male-Male Aggressive Behavior Is Changed by Body-Size Difference in the Leaf-Footed Plant Bug, *Leptoglossus australis*, Fabricius (Heteroptera, Coreidae)," *Journal of Ethology* 11 (1993): 63–65; Miyatake, "Functional Morphology of the Hind Legs as Weapons for Male Contests in *Leptoglossus australis* (Heteroptera: Coreidae)," *Journal of Insect Behavior* 10 (1997): 727–35; W. G. Eberhard, "Sexual Behavior of *Acanthocephala declivis guatemalana* (Hemiptera: Coreidae) and the Allometric Scaling of Their Modified Hind Legs," *Annals of the Entomological Society of America* 91 (1998): 863–71.
12. P. Bergeron, S. Grignolio, M. Apollonio, B. Shipley, and M. Festa-Bianchet, "Secondary Sexual Characters Signal Fighting Ability and Determine Social Rank in Alpine Ibex (*Capra ibex*)," *Behavioral Ecology and Sociobiology* 64 (2010): 1299–307.
13. C. Barrette and D. Vandal, "Sparring, Relative Antler Size, and Assessment in Male Caribou," *Behavioral Ecology and Sociobiology* 26 (1990): 383–87.
14. The "paradox of peace" is predicted by game theory models of assessment. Theoretically, a perfect cue would result in total peace, since all disputes would be settled conventionally, without battle. See, for example, G. Parker, "Assessment Strategy and the Evolution of Animal Conflicts," *Journal of Theoretical Biology* 47 (1974): 223–43.
15. Here I build this story around costs; specifically, the observation that costs are steeper for individuals with fewer resources. This idea is an integral assumption of most models of animal signaling, and it surely applies most of the time. But there are exceptions. As I was writing this book, a doctoral student in my lab, Erin McCullough, was systematically unraveling this notion for the rhinoceros beetle we study. Her work shook the field, because everyone—including myself—assumed that the giant pitchfork horns in these beetles were

costly. How could they not be? These weapons are two-thirds the length of the beetle, and they splay forward in front of the animal's face like a massive pitchfork. Yet, these horns turned out to be remarkably inexpensive to produce, and virtually cost-free to fly around with. For readers interested her work, see E. L. McCullough, P. R. Weingarden, and D. J. Emlen, "Costs of Elaborate Weapons in a Rhinoceros Beetle: How Difficult Is It to Fly with a Big Horn?" *Behavioral Ecology* 23 (2012): 1042–48; E. L. McCullough and B. W. Tobalske, "Elaborate Horns in a Giant Rhinoceros Beetle Incur Negligible Aerodynamic Costs," *Proceedings of the Royal Society of London, Series B* 280 (2013): 1–5; and E. L. McCullough and D. J. Emlen, "Evaluating Costs of a Sexually Selected Weapon: Big Horns at a Small Price," *Animal Behaviour* 86 (2013) 977–85.

16. Most observers don't count these early stages in their tallies because they dissipate so quickly. Barrette and Vandal did include them in their two-year study of caribou contests. Of the 11,640 male-male interactions they observed, 10,332 ended at this initial stage. Barrette and Vandal, "Sparring, Relative Antler Size, and Assessment in Male Caribou," 383–87.
17. A. Berglund, A. Bisazza, and A. Pilastro, "Armaments and Ornaments: An Evolutionary Explanation of Traits of Dual Utility," *Biological Journal of the Linnean Society* 58 (1996): 385–99.
18. D. S. Pope, "Testing Function of Fiddler Crab Claw Waving by Manipulating Social Context," *Behavioral Ecology and Sociobiology* 47 (2000): 432–37; M. Murai and P. R. Y. Backwell, "A Conspicuous Courtship Signal in the Fiddler Crab *Uca perplexa*: Female Choice Based on Display Structure," *Behavioral Ecology and Sociobiology* 60 (2006): 736–41; D. K. McLain and A. E. Pratt, "Approach of Females to Magnified Reflections Indicates That Claw Size of Waving Fiddler Crabs Correlates with Signaling Effectiveness," *Journal of Experimental Marine Biology and Ecology* 343 (2007): 227–38.
19. T. Detto, "The Fiddler Crab *Uca mjoebergi* Uses Colour Vision in Mate Choice," *Proceedings of the Royal Society, Series B* 274 (2007): 2785–90.
20. Dietrich Burkhardt and Ingrid de la Motte, "Big 'Antlers' are Favoured: Female Choice in Stalk-Eyed Flies (Diptera, Insecta), Field Collected Harems and Laboratory Experiments," *Journal of Comparitive Physiology A* 162 (1988): 649–52; G. S. Wilkinson and P. R. Reillo, "Female Choice Response to Artificial Selection on an Exaggerated Male Trait in a Stalk-Eyed Fly," *Proceedings of the Royal Society of London, Series B* 255 (1994): 1–6; G. S. Wilkinson, H. Kahler, and R. H. Baker, "Evolution of Female Mating Preferences in Stalk-Eyed Flies," *Behavioral Ecology* 9 (1998): 525–33.
21. A. J. Moore and P. Wilson, "The Evolution of Sexually Dimorphic Earwig Forceps: Social Interactions Among Adults of the Toothed Earwig, *Vostox apicedentatus*," *Behavioral Ecology* 4 (1993): 40–48; J. L. Tomkins and L. W. Simmons, "Female Choice and Manipulations of Forceps Size and Symmetry in the Earwig *Forficula auricularia* L.," *Animal Behaviour* 56 (1998): 347–56.
22. A. Malo, E. R. S. Roldan, J. Garde, A. J. Soler, and M. Gomendio, "Antlers Honestly Advertise Sperm Production and Quality," *Proceedings of the Royal Society of London, Series B* 272 (2005): 149–57.
23. A. Balmford, A. M. Rosser, and S. D. Albon, "Correlates of Female Choice in Resource-Defending Antelope," *Behavioral Ecology and Sociobiology* 31 (1992): 107–14.
24. N. A. M. Rodger, *The Command of the Ocean—A Naval History of Britain 1649–1815* (W. W. Norton, 2005).
25. R. L. O'Connell, *Of Arms and Men: A History of War, Weapons, and Aggression* (Oxford: Oxford University Press, 1989); O'Connell, *Soul of the Sword: An Illustrated History of Weaponry and Warfare from Prehistory to the Present* (New York: Free Press, 2002); R. Gardiner and B. Lavery, *The Line of Battle: The Sailing Warship 1650–1840* (London: Conway Maritime Press, 2004).
26. Gardiner and Lavery, *Line of Battle.*
27. O'Connell, *Of Arms and Men*; O'Connell, *Soul of the Sword.*

28. Ibid.
29. Gardiner and Lavery, *Line of Battle.*
30. Ibid.
31. D. Miller and L. Peacock, *Carriers: The Men and the Machines* (New York: Salamander Press, 1991).
32. *Wikipedia, s.v.* "Boeing F/A-18E/F Super Hornet."
33. Miller and Peacock, *Carriers.*

10. Sneaks and Cheats

1. D. J. Emlen, "Alternative Reproductive Tactics and Male-Dimorphism in the Horned Beetle *Onthophagus acuminatus* (Coleoptera: Scarabaeidae)," *Behavioral Ecology and Sociobiology* 41 (1997): 335–41.
2. Ibid.
3. The fact that small males "switch off" horn growth provided an unusual opportunity to study developmental mechanisms associated with horn development. Genetically similar males could be raised under conditions that either triggered or suppressed horn growth, and these animals could then be compared in terms of hormone levels, patterns of cell growth, and gene expression. Like a "toe in the door," horn dimorphism afforded rare glimpses into details of insect development. For papers examining how developmental hormones appear to regulate horn growth, see D. J. Emlen and H. F. Nijhout, "Hormonal Control of Male Horn Length Dimorphism in the Dung Beetle *Onthophagus taurus* (Coleoptera: Scarabaeidae)," *Journal of Insect Physiology* 45 (1999): 45–53; D. J. Emlen and H. F. Nijhout, "Hormonal Control of Male Horn Length Dimorphism in *Onthophagus taurus* (Coleoptera: Scarabaeidae): A Second Critical Period of Sensitivity to Juvenile Hormone," *Journal of Insect Physiology* 47 (2001): 1045–54; and A. P. Moczek and H. F. Nijhout, "Developmental Mechanisms of Threshold Evolution in a Polyphenic Beetle," *Evolution and Development* 4 (2002): 252–64. For a comparative study examining the evolution of horn dimorphism in dung beetles, see D. J. Emlen, J. Hunt, and L. W. Simmons, "Evolution of Sexual Dimorphism and Male Dimorphism in the Expression of Beetle Horns: Phylogenetic Evidence for Modularity, Evolutionary Lability, and Constraint," *American Naturalist* 166 (2005): S42–S68; and for more recent studies examining patterns of gene expression in developing horns, see A. P. Moczek and L. M. Nagy, "Diverse Developmental Mechanisms Contribute to Different Levels of Diversity in Horned Beetles," *Evolution and Development* 7 (2005): 175–85; A. P. Moczek and D. J. Rose, "Differential Recruitment of Limb Patterning Genes During Development and Diversification of Beetle Horns," *Proceedings of the National Academy of Sciences* 106 (2009): 8992–97; T. Kijimoto, J. Costello, Z. Tang, A. P. Moczek, and J. Andrews, "EST and Microarray Analysis of Horn Development in *Onthophagus* Beetles," *BMC Genomics* 10 (2009): 504; E. C. Snell-Rood, A. Cash, M. V. Han, T. Kijimoto, J. Andrews, and A. P. Moczek, "Developmental Decoupling of Alternative Phenotypes: Insights From the Transcriptomes of Horn-Polyphenic Beetles," *Evolution* 65 (2011): 231–45.
4. A. P. Moczek and D. J. Emlen, "Male Horn Dimorphism in the Scarab Beetle, *Onthophagus taurus*: Do Alternative Reproductive Tactics Favour Alternative Phenotypes?" *Animal Behaviour* 59 (2000): 459–66; R. Madewell and A. P. Moczek, "Horn Possession Reduces Maneuverability in the Horn-Polyphenic Beetle, *Onthophagus nigriventris*," *Journal of Insect Science* 6 (2006): 21.
5. L. W. Simmons, J. L. Tomkins, and J. Hunt, "Sperm Competition Games Played by Dimorphic Male Beetles," *Proceedings of the Royal Society of London, Series B* 266 (1999): 145–50.
6. For an overview of alternative reproductive tactics in animals, see R. F. Oliveira, M. Taborsky, and H. J. Brockmann, eds., *Alternative Reproductive Tactics: An Integrative Approach* (Cambridge: Cambridge University Press, 2008).
7. J. T. Hogg and S. H. Forbes, "Mating in Bighorn Sheep: Frequent Male Reproduction via a

High-Risk 'Unconventional' Tactic," *Behavioral Ecology and Sociobiology* 41 (1997): 33–48; D. W. Coltman, M. Festa-Bianchet, J. T. Jorgenson, and C. Strobeck, "Age-Dependent Sexual Selection in Bighorn Rams," *Proceedings of the Royal Society of London, Series B* 269 (2002): 165–72.

8. M. R. Gross and E. L. Charnov, "Alternative Male Life Histories in Bluegill Sunfish," *Proceedings of the National Academy of Sciences* 77 (1980): 6937–40; W. J. Dominey, "Maintenance of Female Mimicry as a Reproductive Strategy in Bluegill Sunfish (*Lepomis macrochirus*)," *Environmental Biology of Fishes* 6 (1981): 59–64; M. R. Gross, "Disruptive Selection for Alternative Life Histories in Salmon," *Nature* 313 (1985): 47–48; C. J. Foote, G. S. Brown, and C. C. Wood, "Spawning Success of Males Using Alternative Mating Tactics in Sockeye Salmon, *Oncorhynchus nerka*," *Canadian Journal of Fisheries and Aquatic Sciences* 54 (1997): 1785–95.
9. J. G. van Rhijn, "On the Maintenance and Origin of Alternative Strategies in the Ruff *Philomachus pugnax*," *Ibis* 125 (1983): 482–98; D. B. Lank, C. M. Smith, O. Hanotte, T. Burke, and F. Cooke, "Genetic Polymorphism for Alternative Mating Behaviour in Lekking Male Ruff *Philomachus pugnax*," *Nature* 378 (1995): 59–62.
10. J. Jukema and T. Piersma, "Permanent Female Mimics in a Lekking Shorebird," *Biology Letters* 2 (2006): 161–64.
11. Ibid.
12. S. M. Shuster and M. J. Wade, "Female Copying and Sexual Selection in a Marine Isopod Crustacean, *Paracerceis sculpta*," *Animal Behaviour* 41 (1991): 1071–78; S. M. Shuster, "The Reproductive Behaviour of α-, β-, and γ-Male Morphs in *Paracerceis sculpta*, a Marine Isopod Crustacean," *Behaviour* 121 (1992): 231–58; S. M. Shuster and M. J. Wade, "Equal Mating Success Among Male Reproductive Strategies in a Marine Isopod," *Nature* 350 (1991): 608–10.
13. Ibid.
14. R. T. Hanlon, M.-J. Naud, P. W. Shaw, J. T. Havenhand, "Behavioural Ecology: Transient Sexual Mimicry Leads to Fertilization," *Nature* 433 (2005): 212.
15. Ibid.
16. Sun Tzu, *The Art of War*, trans. Samuel B. Griffith (New York: Oxford University, 1963); Mark McNeilly, *Sun Tzu and the Art of Modern Warfare* (Oxford: Oxford University Press, 2001).
17. Andrew Mack, "Why Big Nations Lose Small Wars: The Politics of Asymmetric Conflict," *World Politics* 27 (1975): 175–200; Ivan Arreguin-Toft, "How the Weak Win Wars: A Theory of Asymmetric Conflict," *International Security* 26 (2001): 93–128.
18. Ibid.
19. Ibid.
20. Raphael Perl and Ronald O'Rourke, "Terrorist Attack on USS *Cole*: Background and Issues for Congress," in *Emerging Technologies: Recommendations for Counter-Terrorism*, eds. Joseph Rosen and Charles Lucey (Hanover, NH: Institute for Security Technology Studies, Dartmouth University, 2001): 52–58.
21. Trevor N. Dupuy, *The Evolution of Weapons and Warfare* (New York: Dacapo Press, 1984).
22. Brian Mazanec, "The Art of (Cyber) War," *Journal of International Security Affairs* 16 (2009): 3–19; Jason Fritz, "How China Will Use Cyber Warfare to Leapfrog in Military Competitiveness," *Culture Mandala: The Bulletin of the Centre for East-West Cultural and Economic Studies* 8 (2008): 28–80.
23. Ibid.
24. Mark Clayton, "Chinese Cyberattacks Hit Key US Weapons Systems: Are They Still Reliable?" *Christian Science Monitor*, May 28, 2013; Ewen MacAskill, "Obama to Confront Chinese President Over Spate of Cyber-Attacks on US," *Guardian*, May 28, 2013; Fritz, "How China Will Use Cyber Warfare to Leapfrog in Military Competitiveness," 28–80.
25. Ibid.

26. Leyla Bilge and Tudor Dumitras, "Before We Knew It: An Empirical Study of Zero-Day Attacks in the Real World," *Proceedings of the 2012 ACM Conference on Computer and Communications Security*, 2012: 833–44.

11. End of the Race

1. F. Kottenkamp, *History of Chivalry and Ancient Armour* (London: Willis and Sotheran Publishers, 1857); G. Duby, *The Chivalrous Society*, trans. Cynthia Poston (Berkeley: University of California Press, 1977); R. L. O'Connell, *Of Arms and Men: A History of War, Weapons, and Aggression* (Oxford: Oxford University Press, 1989); J. France, *Western Warfare in the Age of the Crusades, 1000-1300* (Ithaca, NY: Cornell University Press, 1999).
2. Trevor N. Dupuy, *The Evolution of Weapons and Warfare* (New York: Da Capo Press, 1984).
3. Ibid.
4. Ibid.
5. Dupuy, *Evolution of Weapons and Warfare*; O'Connell, *Of Arms and Men*; R. L. O'Connell, *Soul of the Sword: An Illustrated History of Weaponry and Warfare from Prehistory to the Present* (New York: The Free Press, 2002).
6. Ibid.
7. Ibid.
8. Ibid.
9. Ibid.
10. Trevor Dupuy provides a great overview of the battle of Crécy in his book *The Evolution of Weapons and Warfare*. For more comprehensive treatments of the battle see Henri de Wailly, *Crécy 1346: Anatomy of a Battle* (Poole, NY: Blandford Press, 1987); and A. Ayton, P. Preston, F. Autrand, and B. Schnerb, *The Battle of Crécy, 1346* (Woodbridge, Suffolk, UK: Boydell Press, 2005).
11. Ibid.
12. Ibid.
13. A vivid description of the soldier's perspective of the battle of Agincourt is provided by John Keegan in his book *The Face of Battle: A study of Agincourt, Waterloo, and the Somme* (London: Penguin, 1983). For more comprehensive coverage of this battle, see J. Barker, *Agincourt: The King, the Campaign, the Battle* (London: Little Brown, 2005); and A. Curry, *Agincourt: A New History* (London: Tempus Publishing, 2005).
14. Dupuy, *Evolution of Weapons and Warfare*; O'Connell, *Of Arms and Men*; O'Connell, *Soul of the Sword*.
15. Surprisingly few studies have actually measured the strength and nature of selection on extreme weapons—it's incredibly labor-intensive to do. But those that have often find that as weapons get bigger and bigger the success of males rises up to a point, beyond which success begins to drop. Males with the very largest weapons tend to do worse than males with slightly smaller weapons. If males with the biggest weapons of all fared the best, selection would be "open-ended" and directional. The fact that the very biggest males did slightly worse shows us that selection in these populations is stabilizing, and it suggests these populations may be at or close to their balance point. Examples of this type of selection can be found in harlequin beetles, amphipod crustaceans, and red deer. D. W. Zeh, J. A. Zeh, and G. Tavakilian, "Sexual Selection and Sexual Dimorphism in the Harlequin Beetle *Acrocinus longimanus*," *Biotropica* (2002): 86–96; G. A. Wellborn, "Selection on a Sexually Dimorphic Trait in Ecotypes Within the *Hyalella azteca* Species Complex (Amphipoda: Hyalellidae)," *American Midland Naturalist* 143 (2000): 212–25; L. E. B. Kruuk, J. Slate, J. M. Pemberton, S. Brotherstone, F. Guinness, and T. Clutton-Brock, "Antler Size in Red Deer: Heritability and Selection but no Evolution," *Evolution* 56 (2002): 1683–95.
16. To use the same basic equation I allude to in the text, the benefits of having big weapons (B), representing the offspring produced by the females they guard, are offset by production,

maintenance, and fighting costs (C). When B – C > 0, then selection drives the evolution of weapons toward ever greater size. What's not shown is our assumption that males are gleaning 100 percent of the fertilizations of the females they defend: (1 * B) – C > 0. Cheating males erode the rewards to dominant males, so that they reap only a fraction of the sired offspring that they otherwise would have. In the hypothetical beetle example where sneak males, on average, steal one-fourth of the fertilizations, the new equation would be: (0.75 * B) – C > 0. Benefits are devalued by the proportion of fertilizations stolen by cheaters (1 – 0.25 = 0.75). At some point, if cheaters steal enough, the realized benefits may be low enough that they no longer exceed the costs, even for the most successful guarding males.

17. R. Moen, J. Pastor, and Y. Cohen, "Antler Growth and Extinction of Irish Elk," *Evolutionary Ecology Research* 1 (1999): 235–49.
18. Ibid.
19. R. Baker and G. Wilkinson, "Phylogenetic Analysis of Sexual Dimorphism and Eye Stalk Allometry in Stalk-Eyed Flies (Diopsidae)," *Evolution* 55 (2001): 1373–85; M. Kotrba, "Baltic Amber Fossils Reveal Early Evolution of Sexual Dimorphism in Stalk-Eyed Flies (Diptera: Diopsidae)," *Organisms, Diversity and Evolution* 2004 (2004): 265–75.
20. T. Hosoya and K. Araya, "Phylogeny of Japanese Stag Beetles (Coleoptera: Lucanidae) Inferred from 16s mtrRNA Gene Sequences, with References to the Evolution of Sexual Dimorphism of Mandibles," *Zoological Science* 22 (2005): 1305–18.
21. M. Tabana and N. Okuda, "Notes on *Nicagus japonicus* Nagel," *Gekkan-Mushi* 292 (1992): 17–21; K. Katovich and N. L. Kriska, "Description of the Larva of *Nicagus obscurus* (LeConte) (Coleoptera: Lucanidae: Nicaginae), with Comments on Its Position in Lucanidae and Notes on Adult Habitat," *Coleopterists Bulletin* 56 (2002): 253–58.
22. D. J. Emlen, J. Marangelo, B. Ball, and C. W. Cunningham, "Diversity in the Weapons of Sexual Selection: Horn Evolution in the Beetle Genus *Onthophagus* (Coleoptera: Scarabaeidae)," *Evolution* 59 (2005): 1060-84.
23. T. M. Caro, C. M. Graham, C. J. Stoner, and M. M. Flores, "Correlates of Horn and Antler Shape in Bovids and Cervids," *Behaviorial Ecology and Sociobiology* 55 (2003): 32–41; J. Bro-Jørgensen, "The Intensity of Sexual Selection Predicts Weapon Size in Male Bovids," *Evolution* 61 (2007): 1316–26.
24. J. L. Coggeshall, "The Fireship and Its Role in the Royal Navy" (dissertation, Texas A&M University, 1997); P. Kirsch, *Fireship: The Terror Weapon of the Age of Sail*, trans. John Harland (Barnsley, UK: Seaforth Publishing, 2009).
25. O'Connell, *Of Arms and Men*; Coggeshall, "The Fireship and Its Role in the Royal Navy"; O'Connell, *Soul of the Sword*; R. Gardiner and B. Lavery, *The Line of Battle: The Sailing Warship 1650–1840* (London Conway Maritime Press, 2004); Kirsch, *Fireship.*
26. Ibid.
27. James Coggeshall provides a nice description of the role of fireships in the English attack on the Spanish Armada in his dissertation "The Fireship and Its Role in the Royal Navy." More thorough treatments of this battle are provided in Michael Lewis, *The Spanish Armada* (New York: T. Y. Crowell, 1968); Colin Martin and Geoffrey Parker, *The Spanish Armada* (New York: Penguin Books, 1999).
28. O'Connell, *Of Arms and Men*; Coggeshall, "The Fireship and Its Role in the Royal Navy"; O'Connell, *Soul of the Sword*; Gardiner and Lavery, *Line of Battle*; Robert Jackson, *Sea Warfare: From World War I to the Present* (San Diego: Thunder Bay Press, 2008).
29. Gardiner and Lavery, *Line of Battle.*
30. O'Connell, *Of Arms and Men*; O'Connell, *Sacred Vessels: The Cult of the Battleship and the Rise of the U.S. Navy* (Oxford: Oxford University Press, 1991); O'Connell, *Soul of the Sword*; Gardiner and Lavery, *Line of Battle*; Jackson, *Sea Warfare.*
31. O'Connell, *Of Arms and Men*; O'Connell, *Sacred Vessels*; O'Connell, *Soul of the Sword*; R. K. Massie, *Dreadnought: Britain, Germany, and the Coming of the Great War* (New York: Random House, 2007); Massie, *Castles of Steel: Britain, Germany and the Winning of the Great War at Sea* (New York: Random House, 2008).

32. Ibid.
33. Ibid.
34. Ibid.
35. Ibid.
36. Massie, *Dreadnought*; Massie, *Castles of Steel*; Jackson, *Sea Warfare*.
37. Ibid.
38. Jackson, *Sea Warfare*.
39. Ibid.
40. Ibid.
41. Ibid.
42. Tony Bridgland, *Sea Killers in Disguise: The Story of the Q-Ships and Decoy Ships in the First World War* (Annapolis, MD: Naval Institute Press, 1999).
43. Ibid.
44. Jackson, *Sea Warfare*.
45. Ibid.

12. Castles of Sand and Stone

1. For excellent treatments of the biology of African army ants, or driver ants, see W. H. Gotwald Jr., *The Army Ants: The Biology of Social Predation* (Ithaca, NY: Cornell University Press, 1995); and B. Hölldobler and Edward O. Wilson, *The Ants* (Cambridge, MA: Belknap Press of Harvard University Press, 1990). For vivid and highly readable accounts of ant battles, I recommend works by Mark Moffett, including *Adventures Among Ants: A Global Safari with a Cast of Trillions* (Berkeley: University of California Press, 2010), and "Ants and the Art of War," *Scientific American*, December 2011, 84–9.
2. Hölldobler and Wilson, *Ants*.
3. Caspar Schöning and Mark W. Moffett, "Driver Ants Invading a Termite Nest: Why Do the Most Catholic Predators of All Seldom Take This Abundant Prey?" *Biotropica* 39 (2007): 663–67.
4. W. H. Gotwald Jr., "Predatory Behavior and Food Preferences of Driver Ants in Selected African Habitats," *Annals of the Entomological Society of America* 67 (1974): 877–86; Gotwald, *Army Ants*.
5. For comprehensive treatments of the biology of termites, I highly recommend Takuya Abe, David Edward Bignell, and Masahiko Higashi, *Termites: Evolution, Sociality, Symbioses, Ecology* (Boston: Kluwer Academic Publishers, 2000); and David Edward Bignell, Yves Roisin, and Nathan Lo, *The Biology of Termites: A Modern Synthesis* (New York: Springer, 2011).
6. Johanna P. E. C. Darlington, "Populations in Nests of the Termite *Macrotermes subhyalinus* in Kenya," *Insectes Sociaux* 37 (1990): 158–68.
7. Charles Noirot and Johanna P. E. C. Darlington, "Termite Nests: Architecture, Regulation, and Defence," chap. 6 in Abe, Bignell, and Higashi, *Termites*; Judith Korb, "Termite Mound Architecture, from Function to Construction," chap. 13 in Bignell, Roisin, and Lo, *Biology of Termites*.
8. The history of fortifications is described in Sidney Toy, *Castles: Their Construction and History* (New York: Dover, 1984); Martin Brice, *Stronghold: A History of Military Architecture* (New York: Schocken Books, 1985); J. E. Kaufmann and H. W. Kaufmann, *The Medieval Fortress: Castles, Forts, and Walled Cities of the Middle Ages* (Cambridge, MA: Da Capo Press, 2001); Harold Skaarup, *Siegecraft: No Fortress Impregnable* (New York: iUniverse, 2003); Charles Stephenson, *Castles: A History of Fortified Structures, Ancient, Medieval, and Modern* (New York: St. Martin's Press, 2011).
9. Ibid.
10. Ibid.
11. Ibid.
12. Descriptions of Lachish and its siege can be found in R. D. Barnett, "The Siege of Lachish," *Israel Exploration Journal* 8 (1958): 161–64; David Ussishkin, "The 'Lachish Reliefs' and the

City of Lachish," *Israel Exploration Journal* 30 (1980): 174–95; Kelly Devries, Martin J. Dougherty, Iain Dickie, Phyllis G. Jestice, and Rob S. Rice, *Battles of the Ancient World, 1285 BC–AD 451, from Kadesh to Catalaunian Field* (New York: Metro Books, 2007); Smithsonian, *Military History: The Definitive Visual Guide to Objects of Warfare* (New York: DK Press, 2012).

13. Ibid.
14. Ibid.
15. Descriptions of the Assyrian army and its siege tactics are provided by Devries et al., *Battles of the Ancient World*; Smithsonian, *Military History*.
16. Ibid.
17. Devries et al., *Battles of the Ancient World*.
18. Devries et al., *Battles of the Ancient World*; Smithsonian, *Military History*.
19. Ibid.
20. Skaarup, *Siegecraft*; Devries et al., *Battles of the Ancient World*; Stephenson, *Castles*; Smithsonian, *Military History*.
21. Stephenson, *Castles*.
22. The logic of Lanchester's linear and square laws is discussed in detail in chapter 6, and Caspar Schöning and Mark W. Moffett refer to the role of tunnels with respect to Lanchester's laws in their paper "Driver Ants Invading a Termite Nest," 663–67.
23. Caspar Schöning and Mark Moffett never did figure out what opened the termite mound that day, but an aardvark is the most likely possibility. Certainly, they are the primary termite predator in these areas capable of breaching the walls of a mound, and walls damaged by aardvarks are vulnerable to invasion by siafu until they are repaired.
24. I'm not the first to make such comparisons. Fun early accounts of similarities in the design of human and animal tools, including weapons, is provided in the treatise by the Reverend J. G. Wood, *Nature's Teachings: Human Invention Anticipated by Nature* (London: William Glaisher, High Holborn, 1903). More rigorous and contemporary contrasts between weapons of animals and humans are made by Robert O'Connell in his superb books *Of Arms and Men: A History of War, Weapons and Aggression* (Oxford: Oxford University Press, 1989), and *Soul of the Sword: An Illustrated History of Weaponry and Warfare from Prehistory to the Present* (New York: The Free Press, 2002).
25. The point I'm trying to make is that replication of antlers is tied to reproduction of elk—bulls successful at breeding pass their alleles on to subsequent generations while those that fail to breed do not. But the details are a bit more complex, because alleles influencing the antlers of offspring come from both parents, not just the bulls. Cows and bulls each carry the full complement of the elk genome. Genes important for antler growth may be silenced in cows (since cows do not grow antlers), but they will still be included in her eggs and passed to offspring. Thus, the antlers of sons will reflect the combination of alleles inherited from both parents.
26. Many authors have discussed the pros and cons of comparing cultural evolution with biological evolution, and I would refer readers to the following sources as, in my opinion, some of the best. The classic reference is Luigi Luca Cavalli-Sforza and Marcus J. Feldman, *Cultural Transmission and Evolution: A Quantitative Approach* (Princeton, NJ: Princeton University Press, 1981). A more recent and comprehensive textbook on this topic is Linda Stone, Paul F. Lurquin, and Luigi Luca Cavalli-Sforza, *Genes, Culture, and Human Evolution: A Synthesis* (Malden, MA: Blackwell, 2006). One of my favorite biologists, John Tyler Bonner, examines the evolution of culture in nonhuman animals in his book *The Evolution of Culture in Animals* (Princeton, NJ: Princeton University Press, 1983). I also recommend Paul C. Mundinger, "Animal Cultures and a General Theory of Cultural Evolution," *Ethology and Sociobiology* 1 (1980): 183–223; Jelmer W. Erkins and Carl P. Lipo, "Cultural Transmission, Copying Errors, and the Generation of Variation in Material Culture and the Archaeological Record," *Journal of Anthropological Archaeology* 24 (2005): 316–34; Ruth Mace and Claire J. Holden, "A Phylogenetic Approach to Cultural Evolution,"

Trends in Ecology and Evolution 20 (2005): 116–21; Ilya Tëmkin and Niles Eldridge, "Phylogenetics and Material Cultural Evolution," *Current Anthropology* 48 (2007): 146–54. Finally, as my personal favorites of a growing class of phylogenetic studies of the evolution of cultural traits, I recommend Thomas E. Currie, Simon J. Greenhill, Russell D. Gray, Toshikazu Hasegawa, and Ruth Mace, "Rise and Fall of Political Complexity in Island South East Asia and the Pacific," *Nature* 467 (2010): 801–4; Jared Diamond and Peter Bellwood, "Farmers and Their Languages: The first Expansions," *Science* 300 (2011): 597–603; Remco Bouckaert, Philippe Lemey, Michael Dunn, Simon J. Greenhill, Alexander V. Alekseyenko, Alexei J. Drummond, Russell D. Gray, Marc A. Suchard, and Quentin D. Atkinson, "Mapping the Origins and Expansion of the Indo-European Language Family," *Science* 337 (2012): 957–60.

27. The Reverend J. G. Wood wrote a delightful book on structures manufactured by animals, *Homes Without Hands: Being a Description of the Habitations of Animals, Classed According to Their Principle of Construction* (New York: D. Appleton, 1866). More recent treatises on animal structures are provided in Karl von Frisch, *Animal Architecture* (New York: Harcourt Press, 1974); Richard Dawkins, *The Extended Phenotype: The Gene as Unit of Selection* (Oxford: Oxford University Press, 1984); Richard Dawkins, *The Extended Phenotype: The Long Reach of the Gene* (Oxford: Oxford University Press, 1999); J. Scott Turner, *The Extended Organism: The Physiology of Animal-Built Structures* (Cambridge, MA: Harvard University Press, 2002); Mike Hansell, *Built by Animals: The Natural History of Animal Architecture* (Oxford: Oxford University Press, 2007).

28. Example papers dealing with "horizontal gene transfer," as it is called, include Y. I. Wolf, I. B. Rogozin, N. V. Grishin, and E. V. Kooni, "Genome Trees and the Tree of Life," *Trends in Genetics* 18 (2002): 472–79; E. Bapteste, Y. Boucher, J. Leigh, and W. F. Doolittle, "Phylogenetic Reconstruction and Lateral Gene Transfer," *Trends in Microbiology* 12 (2004): 406–11; J. O. Anderson, "Lateral Gene Transfer in Eukaryotes," *Cellular and Molecular Life Sciences* 62 (2005): 1182–97; Aaron O. Richardson and Jeffrey D. Palmer, "Horizontal Gene Transfer in Plants," *Journal of Experimental Botany* 58 (2007): 1–9; E. Bapteste and R. M. Burian, "On the Need for Integrative Phylogenomics, and Some Steps Toward Its Creation," *Biology and Philosophy* 25 (2010): 711–36.

29. Daniel N. Frank and Norman R. Pace, "Gastrointestinal Microbiology Enters the Metagenomics Era," *Current Opinion in Gastroenterology* 24 (2008): 4–10.

30. Terrence M. Tumpey, Christopher F. Basler, Patricia V. Aguilar, Hui Zeng, Alicia Solórzano, David E. Swayne, Nancy J. Cox, Jacqueline M. Katz, Jeffery K. Taubenberger, Peter Palese, and Adolfo García-Sastre, "Characterization of the Reconstructed 1918 Spanish Influenza Pandemic Virus," *Science* 310 (2005): 77–80; Gavin J. D. Smith, Dhanasekaran Vijaykrishna, Justin Bahl, Samantha J. Lycett, Michael Worobey, Oliver G. Pybus, Siu Kit Ma, Chung Lam Cheung, Jayna Raghwani, Samir Bhatt, J. S. Malik Peiris, Yi Guan, and Andrew Rambaut, "Origins and Evolutionary Genomics of the 2009 Swine-Origin H1N1 Influenza A Epidemic," *Nature* 459 (2009): 1122–25.

31. Charles Ofria, Chris Adami, and Titus Brown developed a software platform for studying evolutionary biology called "Avida," which has proven to be tremendously informative for both research and education. Avida simulates in silico populations of digital organisms, self-replicating units containing digital "genomes." These genomes are used to build digital "bodies" with properties specified in the code. They also incorporate random errors—digital mutations—from time to time as they replicate their code. Digital organisms then compete in a simulated environment, and populations subsequently evolve. Charles Ofria and Richard Lenski run a digital evolution laboratory at Michigan State University (http://devolab.msu.edu/), and recently Ian Dworkin and one of his students used the Avida platform to provide exciting tests of critical elements of sexual selection theory. Christopher Chandler, Charles Ofria, and Ian Dworkin, "Runaway Sexual Selection Leads to Good Genes," *Evolution* 67 (2012): 110–19. Because the Avida system evolves independently of DNA, it provides a groundbreaking means for validating core principles of evolutionary

biology. Avida also shatters the illusion that DNA is somehow "special" as a means of information transfer fueling evolutionary change.

32. The randomness of mutation has confused many people, since it implies that evolution also is random. (If evolution is random, how can exquisite adaptations be explained?) The trick lies in recognizing the difference between the source of variation—where the raw material necessary for evolution comes from—and what happens to this variation once it is there. Natural selection is anything but random. It's no accident that weapons performing poorly are culled, while those performing well are retained and expanded. Given sufficient time and enough variation to work with, natural selection will push the evolution of weapons in directions that are anything but random. Consequently, new mutations infusing genetic variations into biological systems is random, but the evolution that subsequently unfolds in those populations very often is not.
33. A readable and comprehensive treatise on the history (and evolution) of the assault rifle, including especially the success of the AK-47, is provided by C. J. Chivers in his book *The Gun* (New York: Simon and Schuster, 2011).
34. Ibid.
35. Vernon L. Scarborough, Matthew E. Becher, Jeffrey L. Baker, Garry Harris, and Fred Valdez Jr., "Water and Land Use at the Ancient Maya Community of La Milpa," *Latin American Antiquity* 6 (1995): 98–119; N. Hammond, G. Tourtellot, S. Donaghey, and A. Clarke, "Survey and Excavation at La Milpa, Belize," *Mexicon* 18 (1996): 86–91; Gregory Zaro and Brett Houk, "The Growth and Decline of the Ancient Maya City of La Milpa, Belize: New Data and New Perspectives from the Southern Plazas," *Ancient Mesoamerica* 23 (2012): 143–159.
36. David Webster, "The Not So Peaceful Civilization: A Review of Maya War," *Journal of World Prehistory* 14 (2000): 65–119; Elizabeth Arkush and Charles Stanish, "Interpreting Conflict in the Ancient Andes: Implications for the Archaeology of Warfare," *Current Anthropology* 46 (2005): 3–28; Marisol Cortes Rincon, "A Comparative Study of Fortification Developments Throughout the Maya Region and Implications of Warfare" (dissertation, University of Texas, Austin, 2007).
37. In the steep Andes, many towns used terraces instead of traditional walls, which achieved essentially the same effect. For descriptions of Incan and Mayan fortifications, see David Webster, "Lowland Maya Fortifications," *Proceedings of the American Philosophical Society* 120 (1976): 361–71; H. W. Kaufmann and J. E. Kaufmann, *Fortifications of the Incas, 1200–1531* (Oxford: Osprey Publishing, 2006); Rincon, "Comparative Study of Fortification Developments."
38. I highly recommend the paper by Lawrence H. Keeley, Marisa Fontana, and Russell Quick, "Baffles and Bastions: The Universal Features of Fortifications," *Journal of Archaeological Research* 15 (2007): 55–95, as it provides a review of features of early fortifications and worldwide coverage of early examples. For more detail on the specific examples I mention, see James A. Tuck, *Onondaga Iroquois Prehistory: A Study in Settlement Archaeology* (Syracuse, NY: Syracuse University Press, 1971); Merrick Posnansky and Christopher R. Decorse, "Historical Archaeology in Sub-Saharan Africa—A Review," *Historical Archaeology* 20 (1986): 1–14; G. Connah, "Contained Communities in Tropical Africa," in *City Walls*, ed. J. Tracy (Cambridge: Cambridge University Press, 2000): 19–45.
39. Descriptions of the evolution of fortifications in response to escalated siege weapons are provided by Toy, *Castles*; Brice, *Stronghold*; Kaufmann and Kaufmann, *Medieval Fortress*; Skaarup, *Siegecraft*; Stephenson, *Castles.*
40. Ibid.
41. Ibid.
42. Ibid.
43. Duncan B. Campbell, *Greek and Roman Siege Machinery 399 BC–AD 363* (Oxford: Osprey Publishing, 2003).
44. Toy, *Castles*; Brice, *Stronghold*; Kaufmann and Kaufmann, *Medieval Fortress*; Skaarup, *Siegecraft*; Stephenson, *Castles.*

45. Skaarup, *Siegecraft.*
46. Toy, *Castles*; Brice, *Stronghold.*
47. Ibid.
48. René Chartrand, *The Forts of Colonial North America: British, Dutch and Swedish Colonies* (Oxford: Osprey Publishing, 2011).
49. Ron Field, *Forts of the American Frontier 1820–91: Central and Northern Plains* (Oxford: Osprey Publishing, 2005).
50. Brice, *Stronghold*; Skaarup, *Siegecraft.*
51. For an overview of the Japanese tunnels carved into the Pacific Islands during WWII, see Gordon L. Rottman, *Japanese Pacific Island Defenses 1941–45* (Oxford: Osprey Publishing, 2003).
52. Mir Bahmanyar, *Afghanistan Cave Complexes 1979–2004: Mountain Strongholds of the Mujahideen, Taliban, and Al Qaeda* (Oxford: Osprey Publishing, 2004).

13. Ships, Planes, and States

1. John Morrison and John Coates, *Greek and Roman Oared Warships 399–30BC* (Oxford: Oxbow Books, 1997).
2. R. L. O'Connell, *Of Arms and Men: A History of War, Weapons, and Aggression* (Oxford: Oxford University Press, 1989); O'Connell, *Soul of the Sword: An Illustrated History of Weaponry and Warfare from Prehistory to the Present* (New York: The Free Press, 2002).
3. Robert Gardiner, ed., *The Age of the Galley: Mediterranean Oared Vessels Since Pre-Classical Times* (London: Book Sales Publishing, 2000).
4. Lionel Casson, *Ships and Seamanship in the Ancient World* (Baltimore: Johns Hopkins University Press, 1995); Gardiner, *Age of the Galley*; O'Connell, *Soul of the Sword.*
5. Ibid.
6. Trevor N. Dupuy, *The Evolution of Weapons and Warfare* (New York: Da Capo Press, 1984); R. Gardiner and B. Lavery, *The Line of Battle: The Sailing Warship 1650–1840* (London: Conway Maritime Press, 2004).
7. Ibid.
8. Dupuy, *Evolution of Weapons and Warfare*; O'Connell, *Of Arms and Men*; O'Connell, *Soul of the Sword*; Gardiner and Lavery, *Line of Battle.*
9. Ibid.
10. Gardiner and Lavery, *Line of Battle.*
11. Dupuy, *Evolution of Weapons and Warfare*; O'Connell, *Of Arms and Men*; O'Connell, *Soul of the Sword*; Gardiner and Lavery, *Line of Battle*; Robert Jackson, *Sea Warfare: From World War I to the Present* (San Diego: Thunder Bay Press, 2008).
12. For excellent accounts of this early period of air warfare, I recommend Ezra Bowen, *Knights of the Air* (Alexandria, VA: Time-Life Books, 1981); Christopher Campbell, *Aces and Aircraft of World War I* (Dorset, UK: Blandford Press, 1981); Christopher Chant, *Warplanes* (London: M. Joseph, 1983); Robert Jackson, *Aerial Combat* (London: Cox and Wyman, 1976); Norman Franks, *Aircraft Versus Aircraft* (New York: Crescent Books, 1986); and John Blake, *Flight: The Five Ages of Aviation* (Leicester, UK: Magna Books, 1987). For a discussion of the evolution of airplanes in the context of dogfighting, including the importance of duels, see Dupuy, *Evolution of Weapons and Warfare*; Franks, *Aircraft Versus Aircraft*; O'Connell, *Of Arms and Men*; O'Connell, *Soul of the Sword*; Michael Clarke, "The Evolution of Military Aviation," *Bridge* 34 (2004): 29–35.
13. Bowen, *Knights of the Air*; Dupuy, *Evolution of Weapons and Warfare*; O'Connell, *Of Arms and Men*; O'Connell, *Soul of the Sword.*
14. Ibid.
15. Ibid.
16. Aerial dogfights during WWI had many parallels with duels between knights of the Middle Ages. Pilots decorated their planes with unique and colorful markings visible to other pilots in the air—being recognized as an individual was more important to them than simply being

affiliated with a squadron or a side—and they kept meticulous tallies of not just how many planes they shot down, but *whom* they shot down. Finally, victory in the air translated into societal recognition and fame. But the quality of the weapons—the planes—was unconnected with reproduction. Planes were purchased by governments, not individual pilots. The type or model of plane signaled little about individual wealth or family status, and nothing about society at the time prevented poor-quality pilots from marrying and reproducing. Here, the units that mattered were the planes themselves. Models more maneuverable than other models, or models that could turn faster or climb higher or simply overtake other planes, got manufactured in greater numbers than older, slower, or clunkier models, and the population of planes got sucked into a race as each side strived to develop planes that could outperform those of the other. Perhaps the most telling evidence that fighter aircraft evolution is unfolding at the level of the plane rather than the pilot is the fact that the newest planes don't even have pilots. Pilots were instrumental in creating the conditions that started this race, but the aircraft arms race appears to be proceeding full tilt without them.

17. Chant, *Warplanes*; Jackson, *Aerial Combat*; Franks, *Aircraft Versus Aircraft*; Blake, *Flight*.
18. Douglas C. Dildy and Warren E. Thompson, *F-86 Sabre vs MiG-15: Korea 1950–53* (Oxford: Osprey Publishing, 2013).
19. Clarke, "Evolution of Military Aviation," 29–35.
20. Ibid.
21. Ibid.; see also Benjamin Gal-Or, *Vectored Propulsion, Supermaneuverability, and Robot Aircraft* (New York: Springer-Verlag, 1990).
22. *Wikipedia, s.v.* "Dogfight."
23. Alan Epstein, "The Role of Size in the Future of Aeronautics," *Bridge* 34 (2004): 17–23.
24. Clarke, "Evolution of Military Aviation," 29–35.
25. For a fantastic and authoritative treatment of the strategic bombing campaign over Europe during WWII, including in-depth descriptions of crew experiences during bombing runs, see Donald L. Miller, *Masters of the Air: America's Bomber Boys Who Fought the Air War Against Nazi Germany* (New York: Simon and Schuster, 2007).
26. O'Connell, *Of Arms and Men*; O'Connell, *Soul of the Sword*.
27. Ibid.
28. *Wikipedia, s.v.* "List of countries by GDP (nominal)"
29. Robert E. Looney and Stephen L. Mehay, "United States Defense Spending: Trends and Analysis," in *The Economics of Defense Spending: An International Survey*, eds. Keith Hartley and Todd Sandler (London: Routledge, 1990); Philip D. Winters, "Discretionary Spending: Prospects and Future," Congressional Research Service, report prepared for Congress (2005): RS-22128; D. Andrew Austin and Mindy R. Levit, "Trends in Discretionary Spending," Congressional Research Service, report prepared for Congress (2010): RL-34424.
30. This analogy works at another level as well. Just as luxury items such as yachts cost poor people more, and big weapons such as crab claws and antlers cost poor-quality males more, so, too, big-ticket weapons cost more to build in poor nations than they do in rich ones. An F-5 fighter costs more to produce in Spain or South Korea, for example, than it does in the United States, and still more for nations such as Ecuador. We can afford to mass-produce the planes in larger volumes, cutting the per-plane production cost enormously. We can afford to support a larger research and development program, giving us a head start on new innovations to fighter technology. We can afford to train a larger workforce skilled in the necessary technologies—everything from pilots to fly the plane to mechanics to repair it. All of these factors mean that rich nations pay an awful lot less per weapon for state-of-the-art things such as aircraft, submarines, missiles, and carriers than poorer nations do, exacerbating the differences between the haves and have-nots. For a discussion of these issues, see Michael Brsoska, "The Impact of Arms Production in the Third World," *Armed Forces and Society* 4 (1989): 507–30.
31. There are a great many books devoted to the Cold War. I particularly recommend R. E.

Powaski, *March to Armageddon: The United States and the Nuclear Arms Race, 1939 to the Present* (Oxford: Oxford University Press, 1987); P. Glynn, *Closing Pandora's Box: Arms Races, Arms Control, and the History of the Cold War* (New York: Basic Books, 1992); R. Rhodes, *Arsenals of Folly: The Making of the Nuclear Arms Race* (New York: Vintage Books, 2008); D. Hoffman, *The Dead Hand: The Untold Story of the Cold War Arms Race and Its Dangerous Legacy* (New York: Doubleday, 2009); and James R. Arnold and Roberta Wiener, eds., *Cold War: The Essential Reference Guide* (Santa Barbara, CA: ABC-CLIO, 2012). In addition, for clear, concise discussions of the arms race, I recommend Dupuy, *Evolution of Weapons and Warfare*; O'Connell, *Of Arms and Men*; and O'Connell, *Soul of the Sword*.

32. O'Connell, *Of Arms and Men*; O'Connell, *Soul of the Sword*.
33. Ibid.; see also Dildy and Thompson, *F-86 Sabre vs MiG-15*.
34. O'Connell, *Of Arms and Men*; O'Connell, *Soul of the Sword*.
35. Ibid.; see also Kenneth Macksey, *Tank Versus Tank: The Illustrated Story of Armored Battlefield Conflict in the Twentieth Century* (New York: Barnes and Noble Books, 1999); and Stephen Hart, ed., *Atlas of Armored Warfare from 1916 to the Present Day* (New York: Metro Books, 2012).
36. Ibid.
37. Dupuy, *Evolution of Weapons and Warfare*; R. E. Powaski, *March to Armageddon: The United States and the Nuclear Arms Race, 1939 to the Present* (New York: Oxford University Press, 1987); O'Connell, *Of Arms and Men*; Glynn, *Closing Pandora's Box*; O'Connell, *Soul of the Sword*; Rhodes, *Arsenals of Folly*; Hoffman, *Dead Hand*; Arnold and Wiener, *Cold War*.
38. Ibid.
39. O'Connell, *Of Arms and Men*; Glynn, *Closing Pandora's Box*; O'Connell, *Soul of the Sword*.
40. Dupuy, *Evolution of Weapons and Warfare*; Powaski, *March to Armageddon*; O'Connell, *Of Arms and Men*; Glynn, *Closing Pandora's Box*; O'Connell, *Soul of the Sword*; Rhodes, *Arsenals of Folly*; Hoffman, *Dead Hand*; Arnold and Wiener, *Cold War*.
41. Ibid.
42. Ibid.
43. O'Connell, *Of Arms and Men*; Glynn, *Closing Pandora's Box*; O'Connell, *Soul of the Sword*.
44. Hoffman, *Dead Hand*.
45. O'Connell, *Of Arms and Men*; Glynn, *Closing Pandora's Box*; O'Connell, *Soul of the Sword*.
46. B. M. Russett, *What Price Vigilance? The Burdens of National Defense* (New Haven, CT: Yale University Press, 1970).
47. O' Connell, *of Arms and Men*; Glynn, *Closing Pandora's Box*; O' Connell, *Soul of the Sword*.
48. For example, Russett, *What Price Vigilance?*; and Paul G. Pierpaoli Jr., "Consequences of the Cold War," in Arnold and Wiener, *Cold War*.
49. Ibid.
50. Ibid.
51. Ibid.; see also O'Connell, *Soul of the Sword*.
52. Ibid.

14. Mass Destruction

1. A number of accounts of this incident, and the events leading up to it, have now been published. I recommend Stephen J. Cimbala, "Year of Maximum Danger? The 1983 'War Scare' and US-Soviet Deterrence," *Journal of Slavic Military Studies* 13 (2000): 1–24; Arnav Manchanda, "When Truth Is Stranger Than Fiction: The Able Archer Incident," *Cold War History* 9 (2009): 111–33; D. Hoffman, *The Dead Hand: The Untold Story of the Cold War Arms Race and Its Dangerous Legacy* (New York: Doubleday, 2009); and especially the Ph.D. dissertation of Andrew Russell Garland, "1983: The Most Dangerous Year" (University of Nevada Las Vegas, 2011). I should also point out that not all scholars agree that this incident came as close as it did to the end; for an alternative view, see Vojtech Mastny, "How Able Was 'Able Archer'?: Nuclear Trigger and Intelligence in Perspective," *Journal of Cold War Studies* 11 (2009): 108–23.

2. Ibid.
3. Ibid.
4. For examples, see Charles A. Kupchan, "Life After Pax Americana," *World Policy Journal* 16 (1999): 20–27; Evan Feigenbaum, "China's Challenge to Pax Americana," *Washington Quarterly* 24 (2001): 31–43.
5. For vivid and frightening coverage of the destructive power of nuclear weapons, and the risk of accidental detonation, I recommend Eric Schlosser's *Command and Control: Nuclear Weapons, the Damascus Accident, and the Illusion of Safety* (New York: The Penguin Press, 2013).
6. Jeanne Guillemin, *Biological Weapons: From the Invention of State-Sponsored Programs to Contemporary Bioterrorism* (New York: Columbia University Press, 2005); Mark Wheelis, Lajos Rózsa, and Malcolm Dando, *Deadly Cultures: Biological Weapons Since 1945* (Cambridge, MA: Harvard University Press, 2006); Hoffman, *Dead Hand.*
7. Ibid.
8. Ibid.

Index

Page numbers in *italics* refer to illustrations.